Noise and Vibration Control - From Theory to Practice

Edited by Ehsan Noroozinejad Farsangi

Published in London, United Kingdom

IntechOpen

Supporting open minds since 2005

Noise and Vibration Control - From Theory to Practice
http://dx.doi.org/10.5772/intechopen.77592
Edited by Ehsan Noroozinejad Farsangi

Contributors
Lihua Yang, Haipeng Zhang, Shuyong Liu, Haifeng Li, Hailong Jiang, Haiping Wu, Cosmas Pagwiwoko, Louis Jézéquel, Anna Pribilova, Jiri Pribil, Ivan Frollo, Ming Yuan, Fan Yang, Masaaki Mori, Kunihiko Ishihara, Toru Yazawa, Shinji Omata, Lanphuong Nguyen, Ilesanmi Afolabi Daniyan, Khumbulani Mpofu

First published in London, United Kingdom, 2019 by IntechOpen
IntechOpen is the global imprint of INTECHOPEN LIMITED, registered in England and Wales, registration number: 11086078, The Shard, 25th floor, 32 London Bridge Street
London, SE19SG - United Kingdom
Printed in Croatia

British Library Cataloguing-in-Publication Data
A catalogue record for this book is available from the British Library

Additional hard and PDF copies can be obtained from orders@intechopen.com

Noise and Vibration Control - From Theory to Practice
Edited by Ehsan Noroozinejad Farsangi
p. cm.
Print ISBN 978-1-78984-770-3
Online ISBN 978-1-78984-771-0
eBook (PDF) ISBN 978-1-83962-272-4

Meet the editor

Dr. Ehsan Noroozinejad Farsangi is currently serving as an assistant professor in the Department of Earthquake Engineering at the Graduate University of Advanced Technology, Iran. Dr. Noroozinejad is the Chief Editor of the *International Journal of Earthquake and Impact Engineering*, the Associate Editor of the *ASCE Practice Periodical on Structural Design and Construction*, the Associate Editor of the *IET Journal of Engineering*, the Associate Editor of *Frontiers in Built Environment: Earthquake Engineering Section*, and the Trial Engineering Editor of *ASCE Natural Hazards Review*. He is also a member of the FIB Commission on Resilient RC Structures, the ASCE Objective Resilience Committee, the RILEM Structural Performance and Design Technical Committee, and also the IEEE P693 Committee on Seismic Design for Substations.

Contents

Preface **XI**

Chapter 1 **1**
Nonlinear Giant Magnetostrictive Actuator and Its Application in Active Control
by Lihua Yang, Haipeng Zhang, Hailong Jiang, Shuyong Liu, Haiping Wu, Haifeng Li and Ehsan Noroozinejad Farsangi

Chapter 2 **15**
Ultrasonic Vibration-Assisted Hot Glass Embossing Process
by Lanphuong Nguyen

Chapter 3 **33**
mDFA Detects Abnormality: From Heartbeat to Material Vibration
by Toru Yazawa and Shinji Omata

Chapter 4 **55**
Countermeasure for High Level Sound Generated from Boiler Tube Bank Duct
by Masaaki Mori and Kunihiko Ishihara

Chapter 5 **81**
Effective Low Frequency Noise Insulation Adopting Active Damping Approaches
by Ming Yuan and Fan Yang

Chapter 6 **97**
Analysis of Energy Relations between Noise and Vibration Produced by a Low-Field MRI Device
by Jiří Přibil, Anna Přibilová and Ivan Frollo

Chapter 7 **119**
Vibration Analysis and Control in the Rail Car System Using PID Controls
by Ilesanmi Afolabi Daniyan and Khumbulani Mpofu

Chapter 8 **137**
Analysing Non-Linear Flutter Vibrations Using System Dynamic Approach
by Cosmas Pandit Pagwiwoko and Louis Jezequel

Preface

Noise and vibration are two interrelated terms in the field of engineering. Vibration is caused by unbalanced inertial forces and moments, whereas noise is the result of such vibrations. Such phenomena have always been a matter of concern for both researchers and practicing engineers. The lower level of vibration can ensure less tear and wear, less tolerance, and longer fatigue life of the machine or structure. This book presents the established fundamentals in the area of passive, semi-active, and active vibration control, and explores the new and emerging technologies and techniques. There has been a considerable amount of effort devoted to the development and realization of methodologies for the control of sound and vibration, and this book covers the latest theoretical, algorithmic, and practical applications, including:

Chapter 1: A hysteresis nonlinear model is established for a giant magnetostrictive actuator to fully describe the actuator dynamic characteristics.

Chapter 2: The dynamic modeling and simulation of proportional–integral–derivative controls for rail car systems are investigated.

Chapter 3: Effective low-frequency noise insulation active damping approaches are illustrated to improve a structure's noise insulation performance.

Chapter 4: The dynamic behavior of aero-elastic vibrations in the presence of non-linear stiffness such as free-play mechanisms and softening or hardening stiffness is investigated.

Chapter 5: Background information on the assistance of ultrasonic vibration in the hot glass embossing process is provided.

Chapter 6: The generation mechanism of the self-sustained tone is clarified experimentally and numerically, and the methods for suppressing a self-sustained tone using baffle plates and perforated plates are investigated.

Chapter 7: The relationship between energy of vibration and noise signals measured in the magnetic resonance imaging scanning area and its vicinity are mapped to minimize these negative impacts.

Chapter 8: Empirical results quantifying various signals from heartbeat to material vibration are investigated.

It is my hope that this book, which integrates the modeling and design aspects of various noise and vibration control techniques, will serve as a comprehensive guide and reference for practicing engineers and educators, and, more importantly, is a welcome-mat for recent graduates entering the vibration control engineering profession.

Dr. Ehsan Noroozinejad Farsangi
Assistant Professor,
Department of Earthquake Engineering,
Graduate University of Advanced Technology,
Kerman, Iran

Chapter 1

Nonlinear Giant Magnetostrictive Actuator and Its Application in Active Control

Lihua Yang, Haipeng Zhang, Hailong Jiang, Shuyong Liu, Haiping Wu, Haifeng Li and Ehsan Noroozinejad Farsangi

Abstract

The giant magnetostrictive actuator has great use in vibration control, but the linear model cannot fully describe its dynamic characteristics. In this chapter, based on the domain wall theory and piezomagnetic theory, a hysteresis nonlinear model is established to fully describe the actuator dynamic characteristics. In combination with the regularisation method, a sliding mode controller has been designed, and the giant magnetostrictive actuator is also studied in the application of active control. Experimental results show that the hysteresis nonlinear model proposed in the chapter can fully describe the actuator's dynamic characteristics in a wider frequency band and the active control also has a much better isolation effect than the passive vibration; it can significantly attenuate the external incentives.

Keywords: giant magnetostrictive actuator, hysteresis nonlinearity, sliding mode control, active control

1. Introduction

Passive vibration isolation has been widely used as an effective isolation method; it can significantly reduce the vibration transmission between mechanical equipment and the base, while the isolation effect is limited in the micro-vibration and low frequencies. Therefore, the active control has been a focus of research at home and abroad [1–4]. With the development of smart materials, intelligent actuators manufactured by those materials including the magneto-rheological actuator, shape memory alloy actuator, giant magnetostrictive actuator (GMA), and others. These actuators have played a huge role in promoting the active control as the executing agency [5, 6].

With the advantages of high-positioning accuracy, fast response, and wide frequency band, among others, the GMA has a wide variety of applications in fields including vibration control and precision positioning [7–10]. Many scholars have studied the linear modelling method of GMA, which is only suitable for describing low-frequency dynamic characteristics [11, 12]. However, the hysteresis nonlinear model based on the domain walls theory can more clearly reveal the coupling relationship among the magnetization process, the stress magnetic machine effect, and stretching amount, and it can more fully describe the GMA's dynamic characteristics on a wide frequency band [13], which is suitable for the active control actuator. Zhang established the GMA dynamics equations and studied the active control on the basis of the proportional-integral-derivative (PID) algorithm, the results showed that the GMA has some damping effect, but the system's adaptive capacity is a little weak [14]. Francesco calculated the actuator's amplitude-frequency curve and studied the active control in a single freedom degree vibration isolation system. The simulations showed the GMA can significantly reduce the force transmitted to the base [15]. Wang designed a magnetostrictive actuator rod and analysed the system structure vibration on the basis of the linear-quadratic regulator (LQR) algorithm, the results showed the GMA can effectively reduce the structure response of acceleration and displacement [16]. However, these studies mainly focus on the linear actuator model; the isolation frequency band is narrow relatively. Therefore, the application of the GMA hysteresis nonlinear model is urgently needed in the active control.

Currently, some of the more commonly active control strategies are the PID control [17], robust control [18], fuzzy control [19], optimal control [20], adaptive control [21], and sliding mode control [22]. In particular, the sliding mode control has no effect on system parameter perturbation and external disturbance when the system is in sliding mode, so it has much better robustness [23], which is more suitable for the active vibration control algorithm.

Based on the domain wall theory [24], this chapter studied the GMA broadband hysteresis nonlinear model and applied it to the active control of the vibration isolation system. By optimising the design of the hysteresis nonlinear model, the actuator linearity is much better than the linear model on a wider frequency band. The results showed that the GMA hysteresis nonlinear model on a wider frequency band can more fully describe its dynamic characteristics. Compared with passive vibration isolation, the active control has a better isolation effect and wider frequency band isolation, and the sliding mode control strategy also has stronger robustness.

2. Hysteresis nonlinear magnetostrictive actuator model

2.1 Magnetostrictive actuator model

The GMA structure is shown in **Figure 1**. It is mainly made of the push rod, preload spring, magnetostrictive rod, drive coil, and bias coil, where the preload mechanism is comprised of the preload spring, push rod, and preload screw. The polarised magnetic field of the bias coil can make the magnetostrictive rod deform in a linear manner and can prevent the multiplier phenomenon from affecting system control precision. When an alternating current is passed through the drive coil, the magnetostrictive rod can also generate an alternating magnetic field, which causes a dynamic redistribution of magnetic domains in the magnetostrictive rod. Then the telescopic length can be obtained microscopically to promote a displacement and thrust the output push rod, which achieves energy conversion by transferring electromagnetic energy into mechanical energy.

Based on the Jiles-Atherton theory [25], many researchers explored the GMA characteristics of the nonlinear hysteresis model, and the structure optimisation and parameter configuration can make the theoretical calculations accurately match with experiments in a higher linearity. Generally, the GMA motion can be equivalent to a single freedom degree model, shown in **Figure 2**. Based on the magnetic domain theory and piezomagnetic theory [26–28], the partial differential equations of the magnetisation process are established. Then the GMA dynamics equation is also deduced with the specific equations as follows:

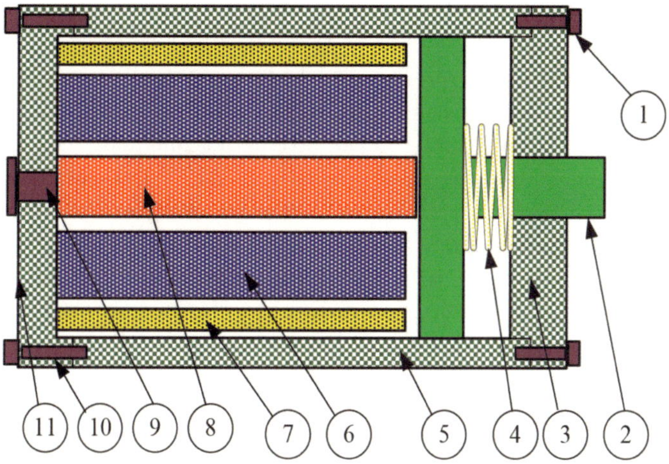

Figure 1.
The GMA structure. (1) The top cover screw, (2) the output push rod, (3) the top cover, (4) the preload spring, (5) the jacket, (6) the drive coil, (7) the bias coil, (8) the magnetostrictive rod, (9) the preload screw, (10) the bottom cover screws, and (11) the bottom cover.

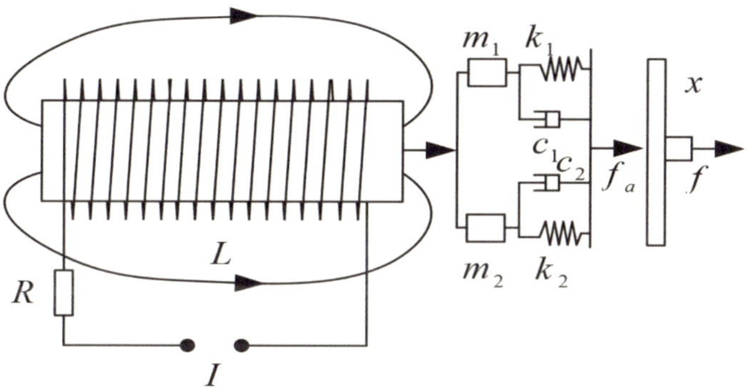

Figure 2.
The GMA schematic diagram.

$$H_e = H + \alpha M + H_\sigma = H + \alpha M + 3\frac{\tilde{\sigma}}{2\mu_0}\left(\frac{d\lambda}{dM}\right)_\sigma \tag{1}$$

$$M_{an} = M_s[\coth(H_e/a) + a/H_e] \tag{2}$$

$$\frac{dM_{irr}}{dH} = \frac{M_{an} - M_{irr}}{\delta k - \tilde{\alpha}(M_{an} - M_{irr})} \tag{3}$$

$$M_{rev} = c(M_{an} - M_{irr}) \tag{4}$$

$$M = M_{irr} + M_{rev} \tag{5}$$

$$\lambda = \frac{3}{2}\frac{\lambda_s}{M_s^2}M^2 \tag{6}$$

$$x_{(s)} = \frac{1}{(\mathrm{m}_1 + m_2)s^2 + (\mathrm{c}_1 + c_2)s + (\mathrm{k}_1 + k_2)}\frac{A}{S^H}\lambda \tag{7}$$

$$f = \frac{m_2 s^2 + c_2 s + k_2}{(m_1 + m_2)s^2 + (c_1 + c_2)s + (k_1 + k_2)}\frac{A}{S^H}\lambda \tag{8}$$

where H_e is the effective magnetic field strength, H is the sum of the drive magnetic field $H_d = nI$ (n is the number of turns per unit coil length, I is the drive current) and bias magnetic field H_b, λ_s is the saturation magnetostrictive coefficient, $\tilde{\sigma} = \sigma_0 + \sigma$ is the external stress (σ_0 is the preload), μ_0 is the vacuum permeability, M_s is the saturation magnetisation, λ is the axial magnetostriction strain, a is the shape factor of non-hysteresis magnetisation, k is the irreversible loss coefficient, α is the molecular field parameter of interacting magnetic torque, $\tilde{\alpha} = \alpha + 9\lambda_s\sigma_0/2\mu_0 M_s^2$ is the integrated magnetic domain coefficient, c is the reversible factor, the parameter $\delta \equiv +1$ when $dH > 0$ and parameter $\delta \equiv -1$ when $dH < 0$, M_{an} is the non-hysteresis magnetisation, M_{irr} is the irreversible magnetisation, M_{rev} is reversible magnetisation, and M is the total magnetisation. C, ρ, L, A, and S^H are, respectively, the damping, density, length and cross-sectional area, and axial compliance coefficient of the magnetostrictive rod; $m_1 = \rho LA/3$, $c_1 = cA/L$, and $k_1 = A/S^H L$ are, respectively, the equivalent mass, damping, and stiffness coefficient of the magnetostrictive rod; m_2, c_2, and k_2 are, respectively, the equivalent mass, damping, and stiffness coefficient of the end load; and s is the Laplace transform operator.

2.2 Magnetostrictive actuator model for wide frequency range

When the driving frequency is low, temperature and eddy current can be ignored on the actuator linearity. However, to fully describe the actuator dynamic characteristics on a wide frequency range, it is necessary to consider the hysteresis loss, eddy current losses, additional loss, and stress changes. In this case, magnetisation intensity M can be expressed as follows:

$$\frac{dM}{dt} = \frac{\partial M}{\partial H}\frac{dH}{dt} + \frac{\partial M}{\partial \sigma}\frac{d\tilde{\sigma}}{dt} \tag{9}$$

where magnetisation change rate is made up of differential susceptibility $\partial M/\partial H$, the magnetic field change rate $\partial H/\partial t$, magnetisation stress change rate $\partial M/\partial \sigma$, and stress change rate $\partial\sigma/\partial t$. The key is to solve for $\partial M/\partial H$ and $\partial\tilde{\sigma}/\partial t$. By Eqs. (2)–(5), the differential susceptibility can be expressed as such:

$$\frac{\partial M}{\partial H} = (1-\mathrm{c})\frac{M_{an} - M_{irr}}{\delta k - \tilde{\alpha}(M_{an} - M_{irr})} + c\frac{\partial M_{an}}{\partial H} \tag{10}$$

According to magneto-mechanical coupling model [16], the magnetisation stress change rate can be determined by Eq. (11):

$$\frac{\partial M}{\partial \sigma} = \frac{S^H \sigma}{\xi}(M_{an} - M) + c\frac{\partial M_{an}}{\partial \sigma} \tag{11}$$

where ξ is the energy coupling parameters of per unit volume and other parameters are the same as above. Substituting Eqs. (10) and (11) into Eq. (9), the magnetisation change rate can be obtained:

$$\begin{aligned}\frac{dM}{dt} = {} & \left((1-\mathrm{c})\frac{M_{an} - M_{irr}}{\delta k - \tilde{\alpha}(M_{an} - M_{irr})} + c\frac{\partial M_{an}}{\partial H}\right)\frac{dH}{dt} \\ & + \left(\frac{S^H \sigma}{\xi}(M_{an} - M) + c\frac{\partial M_{an}}{\partial \sigma}\right)\frac{d\tilde{\sigma}}{dt}\end{aligned} \tag{12}$$

In summary, Eqs. (1)–(12) constitute the hysteresis nonlinear model of the giant magnetostrictive actuator, which can describe the displacement and force output on a wide frequency range, and then the output displacement and force can be calculated by the input current.

3. Giant magnetostrictive actuator experiment

The experiments mainly include the giant magnetostrictive actuator, NI host controller, capture cards and displays, LabVIEW software, FN15150 power amplifier, MEL laser displacement sensor, and CHINT current transformer. NI equipment and software are used to capture and display the current and displacement signal, the amplifier amplifies NI weak signals to drive actuators, and the laser displacement sensor is used to accurately measure the actuator displacement. The GMA experiment setup is shown in **Figure 3**. This experiment studies the hysteresis nonlinear model displacement responses to the preload and the bias magnetic field intensity, as shown in **Figures 4** and **5**. By optimising the actuator preload and bias magnetic field, the GMA can eliminate the doubling phenomenon and work perfectly within a linear range, which can enhance the actuator output linearity.

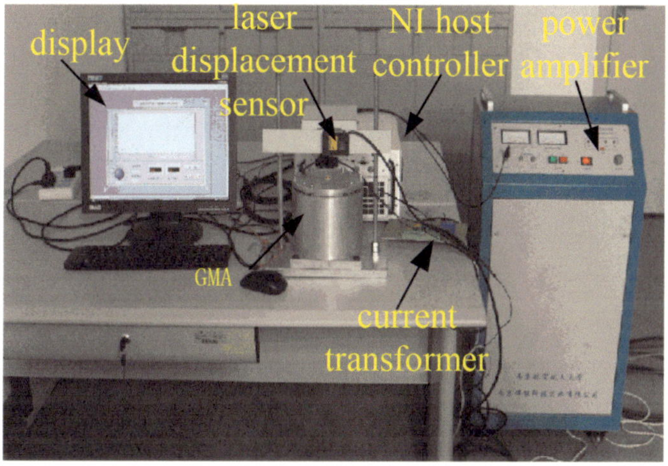

Figure 3.
The giant magnetostrictive actuator experiment.

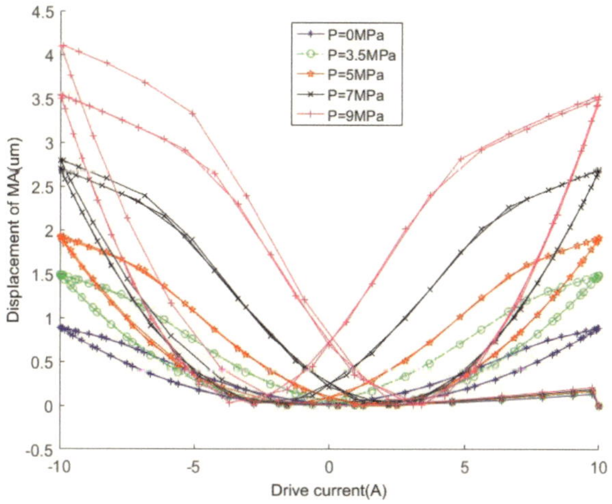

Figure 4.
Curves of displacement changes with the preload.

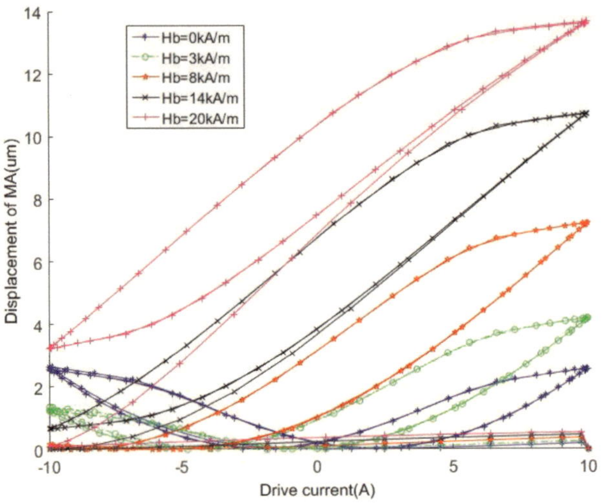

Figure 5.
Curves of displacement changes with the bias magnetic field intensity.

Figure 4 shows that the actuator displacement increases with the preload, but when the preload is greater than 7 MPa, experimental curves began to appear asymmetrical, so the preload cannot be greater than 7 MPa. **Figure 5** shows that without the paranoid magnetic field, actuator output is butterfly-shaped with a strong nonlinearity, but when the paranoid magnetic field increases, the output linearity also gradually increases. Finally, $\sigma_0 = 6.7MPa$ and $H_b = 14kA/m$are selected in the study which can give the actuators better linearity.

4. Application for GMA in active control

4.1 Active control model of vibration isolation system

The active control model of the vibration isolation system is shown in **Figure 6**, where M_1, K_1, C_1, M_2, K_2, and C_2 are, respectively, the mass, stiffness, and damping of the upper isolated equipment and the middle vibration isolation platform. $x, y, \dot{x}$, $\dot{y}, \ddot{x}$, and $\ddot{y}$ are, respectively, displacement, velocity, and acceleration as above. f is the active control force of the GMA, and p is the external disturbance. The dynamics equations of the isolation system can be expressed as Eq. (13):

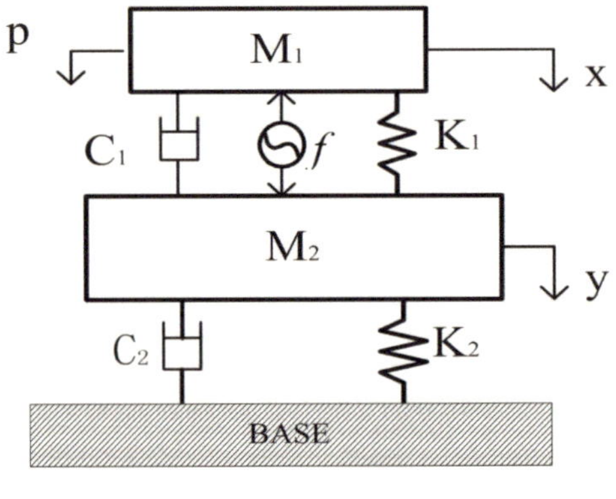

Figure 6.
Active control model of vibration isolation system.

$$\begin{cases} M_1\ddot{x} + C_1(\dot{x} - \dot{y}) + K_1(x - y) = p - f \\ M_2\ddot{y} + C_2\dot{y} + K_2 y + C_1(\dot{y} - \dot{x}) + K_1(y - x) = f \end{cases} \tag{13}$$

Take $X = [x, y, \dot{x}, \dot{y}]^T$ as state variables and the middle platform displacement $Y = y$ as system output. Thus, the state space can be obtained using Eq. (14):

$$\begin{cases} \dot{X}(t) = AX(t) + Bu(t) + B_1 w(t) \\ Y(t) = CX(t) \end{cases} \tag{14}$$

where $A = \begin{bmatrix} 0 & 0 & 1 & 0 \\ 0 & 0 & 0 & 1 \\ -K_1/M_1 & K_1/M_1 & -C_1/M_1 & C_1/M_1 \\ K_1/M_2 & -(K_1+K_2)/M_2 & C_1/M_2 & -(C_1+C_2)/M_2 \end{bmatrix}$,

$B = \begin{bmatrix} 0 \\ 0 \\ -1/M_1 \\ 1/M_2 \end{bmatrix}$, $B_1 = \begin{bmatrix} 0 \\ 0 \\ 1/M_1 \\ 0 \end{bmatrix}$, $C = [0\ 1\ 0\ 0]$, $u = [f]$, $w = [p]$, and $Y = [y]$. Let $U = u + B^{-1}B_1 w$. Eq. (14) can be further expressed as follows:

$$\begin{aligned} \dot{X}(t) &= AX(t) + Bu(t) + B_1 w(t) \\ &= AX(t) + BU(t) \end{aligned} \tag{15}$$

4.2 Sliding mode control design

Based on full state feedback, the sliding mode can make the system reach the sliding mode surface and achieve sliding mode movement in a jump way. Therefore, it is essential for dynamic characteristics to rationally design the sliding mode surface, assuming the system switching function as Eq. (4) [29, 30].

$$S(t) = \Theta X \tag{16}$$

where S is the switching function, Θ is 1×4 dimension switching matrix, and $X = [x, y, \dot{x}, \dot{y}]^T$ are the state variables. The non-singular state transition matrix $\Gamma \in R^{4\times4}$ is taken to regulate Eq. (16), and the coordinate transformation is the following:

$$Z(t) = \Gamma X(t) \tag{17}$$

where $\Gamma = \begin{bmatrix} I_3 & -\tilde{B}_1\tilde{B}_2 \\ 0 & I_1 \end{bmatrix}$, $B = \begin{bmatrix} \tilde{B}_1 \\ \tilde{B}_2 \end{bmatrix}$, $\tilde{B}_1 \in R^{3\times1}$, and $\tilde{B}_2 \in R^{1\times1}$; substituting Eq. (17) into Eqs. (15) and (16), it gets the system canonical form and the switching surface as Eq. (18):

$$\begin{aligned} \dot{Z}(t) &= \overline{A}Z(t) + \overline{B}U(t) \\ S(t) &= \overline{\Theta}Z \end{aligned} \tag{18}$$

where $\overline{A} = \Gamma A \Gamma^{-1}$, $\overline{B} = \begin{bmatrix} 0 & \tilde{B}_2{}^T \end{bmatrix}^T$, and $\overline{\Theta} = \Theta\Gamma^{-1}$; let $Z(t) = \begin{bmatrix} Z_1(t) \\ Z_2(t) \end{bmatrix}$, $\overline{A} = \begin{bmatrix} \overline{A}_{11} & \overline{A}_{12} \\ \overline{A}_{21} & \overline{A}_{22} \end{bmatrix}$, and $\overline{\Theta} = \begin{bmatrix} \overline{\Theta}_1 \\ \overline{\Theta}_2 \end{bmatrix}^T$, where $Z_1(t) \in R^{3\times1}$, $Z_2(t) \in R^{1\times1}$, $\overline{A}_{11} \in R^{3\times3}$, $\overline{A}_{12} \in R^{3\times1}$, $\overline{A}_{21} \in R^{1\times3}$, and $\overline{A}_{22} \in R^{1\times1}$; then Eq. (18) can be decomposed as Eq. (19):

$$\begin{aligned} \dot{Z}_1(\mathrm{t}) &= \overline{A}_{11}Z_1(\mathrm{t}) + \overline{A}_{12}Z_2(\mathrm{t}) \\ S &= \overline{\Theta}_1 Z_1(\mathrm{t}) + \overline{\Theta}_2 Z_2(\mathrm{t}) \end{aligned} \tag{19}$$

Let $S = 0$ and $\overline{\Theta}_2 = I_1$; in Eq. (19), it gets:

$$\begin{aligned} Z_2(\mathrm{t}) &= -\overline{\Theta}_1 Z_1(\mathrm{t}) \\ \dot{Z}_1(\mathrm{t}) &= (\overline{A}_{11} - \overline{A}_{12}\overline{\Theta}_1)Z_1(\mathrm{t}) \end{aligned} \tag{20}$$

where the matrix $\overline{\Theta}_1$ can be designed by the optimal control method or pole assignment method, and then the sliding surface $S(t)$ can also be determined. Finally, the saturation function of exponential reaching law is used as Eq. (21):

$$\dot{S}(t) = -\beta S(t) - \xi sat(S(t)) \qquad \xi > 0, \beta > 0 \tag{21}$$

where $\alpha > 0$ and $1 > \beta > 0$; with the combination of Eqs. (15), (16), and (21), it can get active control force as Eq. (22) by ignoring the external disturbances:

$$f = -(\Theta B)^{-1}[\Theta AX(\mathrm{t}) + \beta S(t) + \xi sat(S(t))] \tag{22}$$

5. Experimental research

In this chapter, compared with passive vibration isolation, the active control experiment is studied to analyse the isolation effect on a double-layer vibration isolation system. The external disturbance is the force of shaker JZK-40, the hardware control system is designed by LabVIEW real time, and the isolation system experiment setup is shown in **Figure 7**.

When the external stimulus is applied to the system, the controllers can capture the acceleration signal of the upper and middle layers; then the signal is conveyed to a controller by an A/D converter. At the same time, D/A signals are given out by NI control calculations, passing through amplifier YE2706A, and the output finally is transmitted to GMA for carrying out the active control. The vibration isolation system control block diagram is shown in **Figure 8**.

In low frequencies, taking the middle platform displacement as the evaluation index, the isolation effect is compared between passive vibration isolation and active control. Experimental parameters are as follows:

$$\begin{aligned} &n = 1200, H_b = 10kA/m,\ a = 7102A/m,\ \tilde{\alpha} = -0.01,\ M_s = 7.65 \times 10^5 A/m, \\ &\lambda_s = 1.005 \times 10^{-6},\ k = 3283A/m,\ b = 0.18,\ S^H = 1.3 \times 10^{-11},\ d = 1.0 \times 10^{-8} m^2/N, \\ &\xi = 7600Pa, C = 3000kNs/m^2,\ \rho = 9250kg/m^3,\ L = 8.6 \times 10^{-4} m,\ A = 78.5 \times 10^{-6} m^2, \\ &M_1 = 155.42kg,\ C_1 = 302.7Ns/m,\ K_1 = 57016.2N/m,\ M_2 = 22.5kg,\ C_2 = 290.544Ns/m, \\ &\text{and } K_2 = 190000N/m. \end{aligned} \tag{23}$$

The system's natural frequency ($f_1 = 2.8Hz, f_2 = 15.6Hz$) can be calculated according to the above parameters, and the external disturbances were taken as single, multi-frequency, and random signals. The experimental results are in **Figures 9–13** and in **Table 1**.

Figures 9–12 show that active control can effectively suppress the vibration generated by external incentives with a significant isolation effect and speed response. **Table 1** shows that, compared with the passive vibration isolation, the

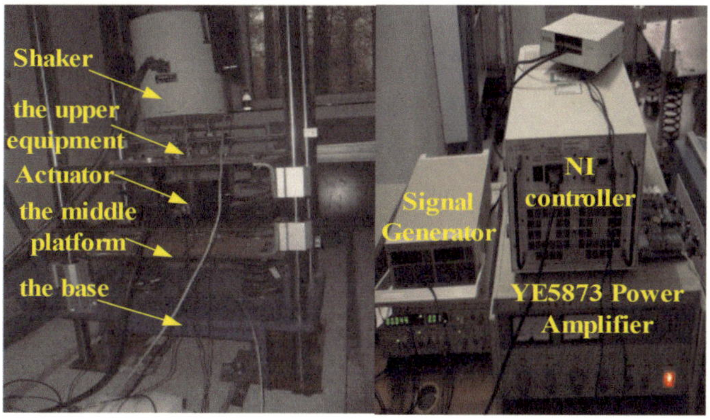

Figure 7.
The isolation system experiment setup.

active control makes the middle raft displacement decrease by 81.08, 78.21, 81.13, and 55.34% in a single frequency, multi-frequency, and random excitation, respectively. Therefore, the active control has obvious advantages and a good isolation effect.

Figure 13 shows that the natural frequencies corresponding to the two peaks of the passive isolation system are fully consistent with the calculations. Moreover, the active control can change the system mode to eliminate the first- and second-order formants to achieve the purpose of isolation. It also shows the active control is poor when the excitation frequency is less than 2 Hz, and then the isolation effect increases with escalating excitation frequency. However, there are always cross-points between active control and passive vibration isolation in the amplitude-frequency curve, which also shows the active control is more suitable for low- and middle-frequency vibration control. It can compensate for the lack of passive vibration isolation and effectively inhibit the transmission of vibration isolation and broaden isolation frequency band, all of which are of great significance for the study of active vibration control in engineering applications.

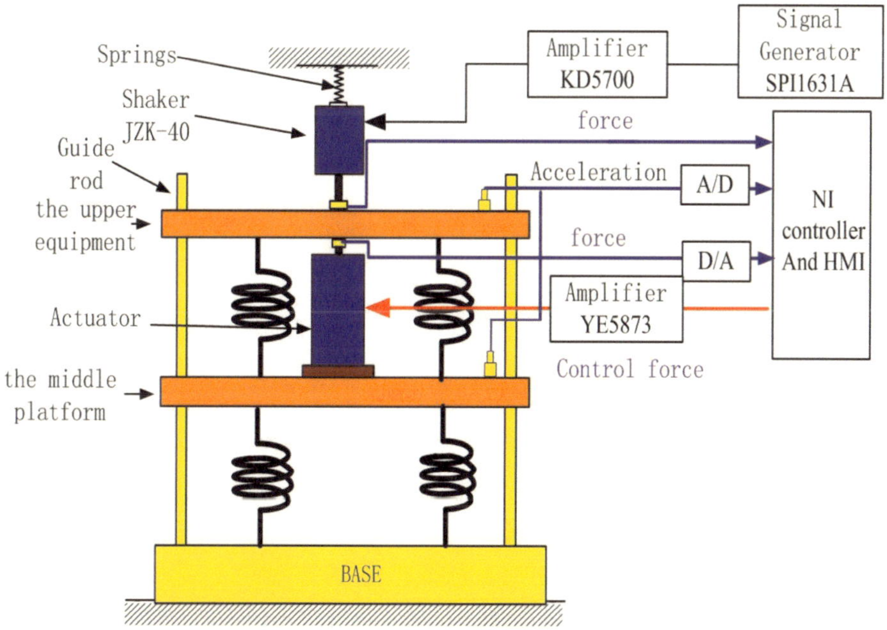

Figure 8.
Vibration isolation system control block diagram.

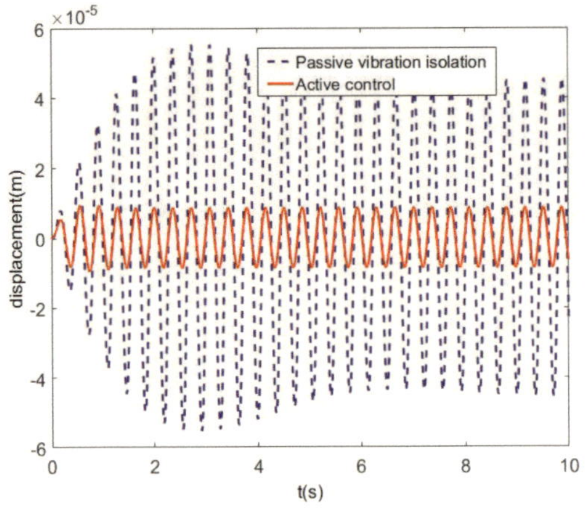

Figure 9.
The time history of middle-layer displacement with f_1 excitation.

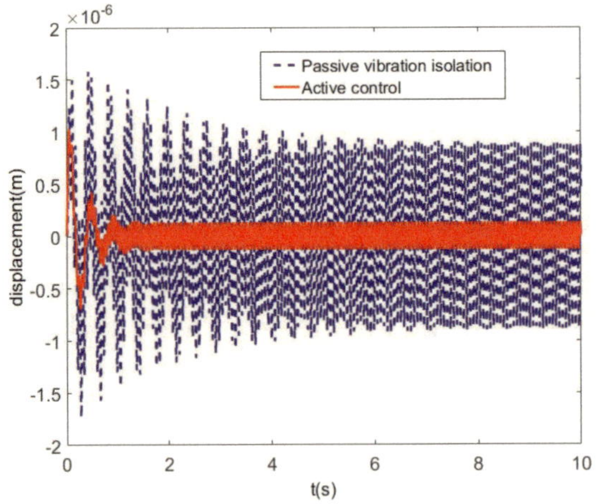

Figure 10.
The time history of middle-layer displacement with f_2 excitation.

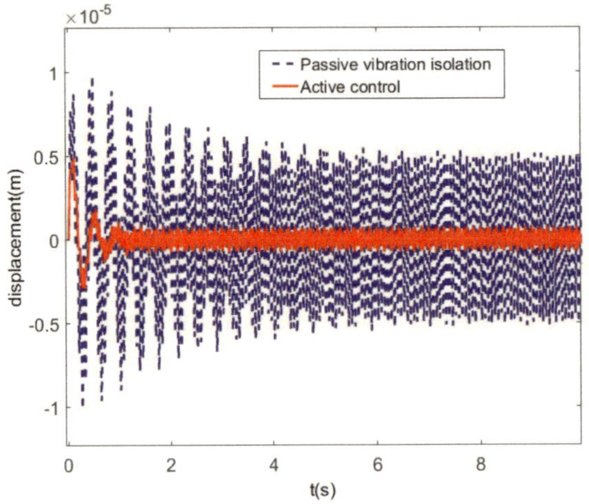

Figure 11.
The time history of middle-layer displacement with $\sqrt{2}f_1 + \sqrt{2}f_2 + 50$ excitation.

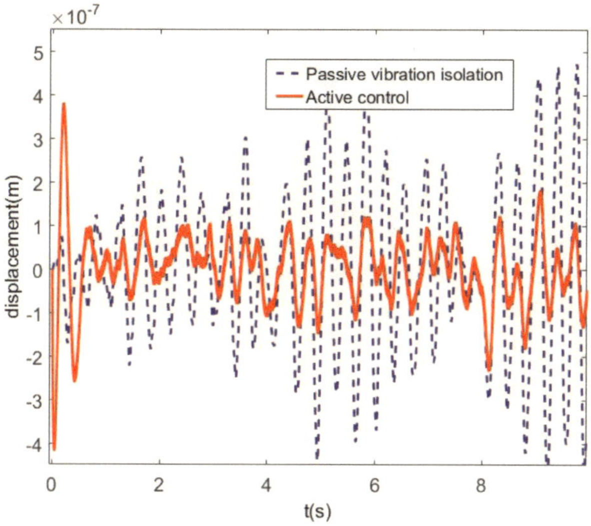

Figure 12.
The time history of middle-layer displacement with random excitation.

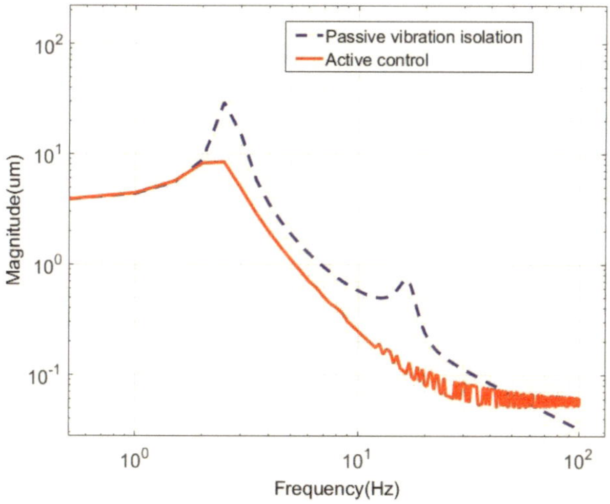

Figure 13.
The amplitude-frequency curves of middle-layer displacement.

	The middle-layer displacement RMS			
	f_1	f_2	$\sqrt{2}f_1 + \sqrt{2}f_2 + 50$	**Random**
Passive vibration isolation	32.1970	0.6576	3.7067	0.1892
Active control	6.0899	0.1432	0.6996	0.0845

Table 1.
The middle-layer displacement RMS with excitations.

6. Conclusions

In this study, GMA linear model cannot fully describe the dynamic behaviour. Based on the magnetic domain theory, piezomagnetic theory, and with the consideration of the magnetic hysteresis, eddy current, and alternating stress, the GMA hysteresis nonlinear model is established, which can describe the actuator dynamic characteristics in a broadband segment. The GMA is then used in active control with the sliding mode control theory. The experimental results show that the hysteresis nonlinear model can fully describe the actuator dynamic characteristics in a broadband segment; the active control is better than passive isolation in single, multi-frequency, and random excitation; and it can effectively broaden the isolation frequency band and improve the response speed and suppress vibration transmission. But it also has some isolation range, which means the effect will deteriorate beyond the effective frequency isolation band.

Acknowledgements

This work was supported by the State Key Laboratory of Ocean Engineering (Shanghai Jiao Tong University) (Grant no. 1714) project ZR2019QEE031 supported by Shandong Provincial Natural Science Foundation and the National Natural Science Foundation of China (51579242, 51509253).

Conflict of interest

The authors declare that there are no conflicts of interest regarding the publication of this chapter.

Author details

Lihua Yang[1*], Haipeng Zhang[1], Hailong Jiang[1], Shuyong Liu[2], Haiping Wu[1], Haifeng Li[1] and Ehsan Noroozinejad Farsangi[3]

1 Power Control Department, Navy Submarine Academy, Qingdao, P.R. China

2 College of Power Engineering, Naval University of Engineering, Wuhan, P.R. China

3 Faculty of Civil Engineering and Surveying, Graduate University of Advanced Technology, Iran

*Address all correspondence to: dreamfly4@163.com

References

[1] Gao HW, Luo J, Jia JY. Modeling and simulation of active control of two-layer vibration isolation system. Mechanical Science and Technology. 2005;**24**(11): 1340-1344

[2] Yang TJ, Gu Z, Liu Z. Experimental research on active control of coupled vibration for a two-stage isolation system. Journal of Vibration Engineering. 2003;**16**(2):25-27

[3] Zhang KG, Du QF. Study on active control of vibration in two-layer isolation system with multiple excitations. China Mechanical Engineering. 2006;**17**(2):134-136

[4] He L, Xu W. Naval vessel machinery mounting technology and its recent advances. Acta Acustica. 2013;**38**(2): 128-135

[5] Wang YF, Cheng W. Modeling and simulation of a mini micro-vibration active control platform for spacecrafts. Journal of Vibration and Shock. 2013; **32**(22):140-145

[6] Quan WF, Mao JQ, Li C. Application of intelligent structure and intelligent control in active vibration control. Information and Electronic Engineering. 2004;**2**(3):232-237

[7] Sathishkumar R, Prasath JS. Terfenol-D: A high power giant magnetostrictive material for submarine mapping. International Journal of Engineering, Science and Technology. 2010;**2**(12):7165-7170

[8] Tao ML, Chen DF, Lu QG. Eddy current losses of giant magnetostrictors: Modeling and experimental analysis. Journal of Mechanical Engineering. 2012;**48**(13):146-151

[9] Hu SF, Zhu SJ. High-precision control of giant magnetostrictive actuator based on CMAC neural network. Journal of Vibration and Shock. 2009;**28**(3):68-72

[10] Lau HY, Liu KP. Feasibility of using GMM based actuators in active control of journal bearing system. In: Proceedings of the World Congress on Engineering; WCE 2009; Vol. II; 1–3 July 2009; London, UK; 2009

[11] Valadkhan S, Morris K, Khajepour A. Review and comparison of hysteresis models for magnetostrictive materials. Journal of Intelligent Material Systems and Structures. 2009;**20**(1):131-142

[12] Jia Z, Wang X, Wang F. Identification method of giant magnetostrictive transducer's dynamic parameters and magnetic parameters. Chinese Journal of Mechanical Engineering. 2007;**43**(10):9-13

[13] Jiles DC, Thoelke JB, Devine MK. Numerical determination of hysteresis parameters for the modeling of magnetic properties using the theory of ferromagnetic hysteresis. IEEE Transactions on Magnetics. 1992;**28**(1): 27-35

[14] Zhang TF, Wang HZ, Sun Y. Simulation on active vibration control. Journal of Vibration and Shock. 2006; **25**(1):61-65

[15] Dezza FC, Cinquemani S. A model of magnetostrictive actuators for active vibration control. In: Proceedings of 2011 IEEE 5th International Conference on Mechanics of Biomaterials & Tissues Italy; 15 February 2011

[16] Wang SL, Ji QB, Dai JB. Study on active vibration control of the structure using giant magnetostrictive actuator. Noise and Vibration Control. 2010;**6**:23-26

[17] Li B, Dong W, Wang X. Design and simulation of an active vibration isolator

based on pneumatic-electromagnetic hybrid driving. Journal of Northwestern Polytechnical University. 2013;**31**(6): 871-877

[18] Hao HR, Bai HB, Zhang HJ. Robust control of a 6-DOF active-passive vibration isolation platform. Journal of Vibration and Shock. 2012;**31**(7):122-127

[19] He X. Fuzzy logic control of double stage vibration isolation system. Noise and Vibration Control. 2010;**2**:38-42

[20] Gao XK, Shao Q. Semi-active optimal control of intelligent damping double vibration isolation system. Journal of Vibration and Shock. 2012; **31**(19):128-133

[21] Li YN, Zhang F, Lei W. Active vibration control of gear meshing based on online secondary path identification algorithm. Journal of Vibration and Shock. 2013;**32**(16):7-12

[22] Qinlei H, Guangfu M, Jiang Y. Variable structure control with time varying sliding mode and vibration control for flexible satellite. Control Theory & Applications. 2009;**26**(2): 122-126

[23] Lan KJ, Yen JY. Sliding mode control for active vibration isolation of a long range scanning tunneling microscope. The Review of Scientific Instruments. 2004;**75**(11):4367-4373

[24] Kronmüller H, Goll D. Micromagnetic theory of the pinning of domain walls at phase boundaries. Physica B. 2002;**319**(1-4):122-126

[25] Jiles DC, Atherton DL. Theory of ferromagnetic hysteresis (invited). Journal of Applied Physics. 1984;**55**(6): 2115-2120

[26] Shu L, Chen DF, Lu QG. Modeling of time-delay in giant magnetostrictive actuator and Smith strategy. Journal of System Simulation. 2009;**21**(10): 3017-3021

[27] Wang B, Cao S, Huang M. Magnetostrictive Materials and Devices. Beijing: Metallurgical Industry Press; 2008

[28] Kim W j, Sadighi A. A novel low-power linear magnetostrictive actuator with local three-phase excitation. IEEE/ASME Transactions on Mechatronics. 2009;**21**(10):3017-3021

[29] Young KD. A control engineer's guide to sliding mode control. IEEE Transactions on Control Systems Technology. 2004;**3**:156-164

[30] Oliveiraa RCLF, de Oliveirab MC, Peresa PLD. Convergent LMI relaxations for robust analysis of uncertain linear systems using lifted polynomial parameter-dependent Lyapunov functions. Systems & Control Letters. 2008;**57**(8):680-668

Chapter 2

Ultrasonic Vibration-Assisted Hot Glass Embossing Process

Lanphuong Nguyen

Abstract

This chapter is intended to provide the reader with background information about the assistance of ultrasonic vibration in hot glass embossing process. For this purpose, the description of the conventional hot glass embossing process and the ultrasonic vibration-assisted hot glass embossing process will be defined first. Based on the comparison between these two processes, components of the ultrasonic vibration device which produces ultrasonic vibration will be discussed in principle. Among these components, ultrasonic horn will be especially analyzed. After that, each interesting effect of ultrasonic vibration on hot glass embossing process will be explained in detail. With the development and application of simulation tools on forming process, this chapter will finally describe some results of finite element analysis (FEA) of ultrasonic vibration-assisted hot glass embossing process.

Keywords: ultrasonic vibration, hot glass embossing, ultrasonic vibration device, microstructures, friction, finite element analysis

1. Introduction

1.1 Conventional hot glass embossing process

Hot embossing process is one of the common replication technologies for the replication of microstructures in micro optic systems, such as micro-lens array, micro-prism array, photoresist columns on a glass substrate, etc. The micro-replication process is known as a process replicated from a microstructured master. With the development of microsystem technology, low cost is one of the important requirements for manufacturing optical components in a micro range. Besides injection molding process, hot embossing has been favored because of its operation as well as the convenience in producing mold tools. Especially, it is the best choice to create micro components with complex geometry and high aspect ratio. In addition, hot embossing is also suitable for mass volume fabrication with high quality.

Material is the main component in molding process, also hot embossing process. Most applicable materials for hot embossing are thermoplastic polymers. Nevertheless, alternative molding materials like glass, metals, or ceramics will also be used [1]. Glass is a material commonly used in optical microsystem. Compared to optical polymers, optical glass has higher transparency, higher scratch, and humid resistance. Another advantage of the optical glass is that its thermal expansion coefficient is smaller than that of the optical polymers. This reduces significantly the deviation between the design and the final optical components. Moreover, the refractive index of the optical glass is much higher than that of optical polymers (a range from 1.5 to over 2.0 compared to a range from 1.3 to 1.7). With the higher refractive index, optical glass could bend the light rays to focus in a smaller range, which is suitable to optical lenses. With the above advantages, optical glass has been favored for high-precision applications.

The hot glass embossing process is divided into four major steps (**Figure 1**). The process starts with heating the sample to the molding temperature, which is usually above transition temperature T_g or the annealing temperature A_t of the material, followed by an isothermal molding by embossing with speed- and/or force-controlled, the cooling of the molded part to the de-molding temperature, and finally de-molding of the component. Heating and embossing stages are usually performed in vacuum environment to protect molds from oxidation, while the cooling stage is usually supported by high-pressure nitrogen. Compared to injection molding, hot embossing process has more advantages, such as shorter flow distances and lower velocity, which decreases shear stress of material significantly. As a result, the decrease of shear stress of material as filling into micro cavities should reduce the residual stress of the embossed parts. In addition, hot embossing process could be the best choice for manufacturing microstructures, which are hardly performed by injection molding. Since the molding temperature is set constantly, the conventional hot glass embossing process is isothermal. The temperature setting of this process is shown in **Figure 2**.

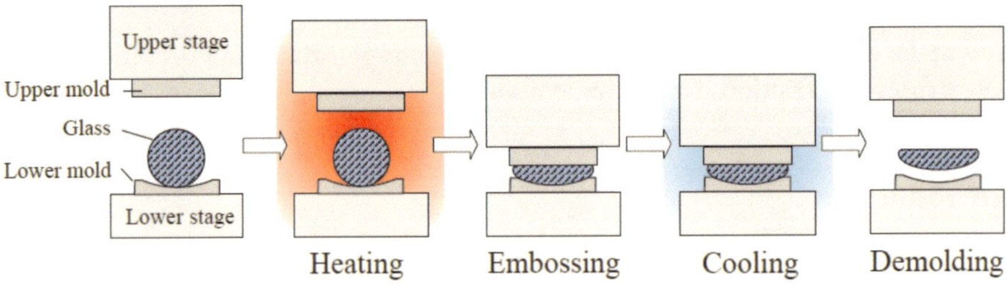

Figure 1.
Steps for the conventional hot glass embossing process [2].

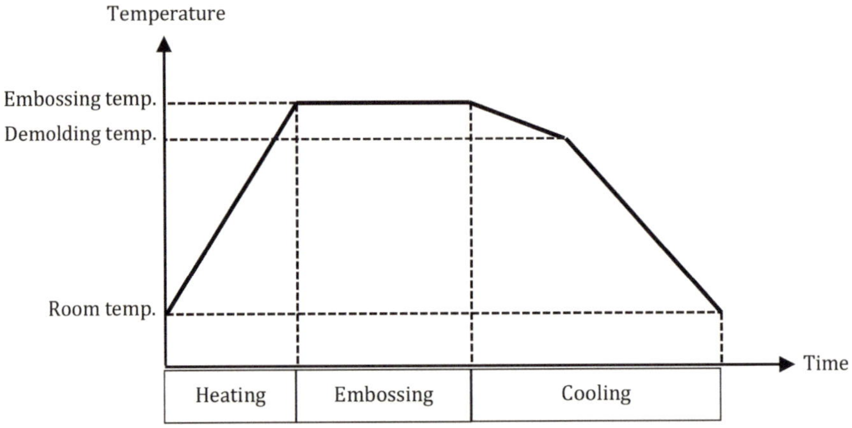

Figure 2.
Temperature setting for the conventional glass hot embossing process.

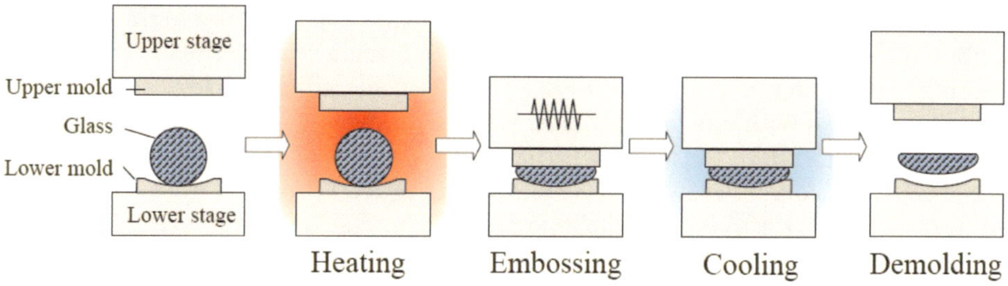

Figure 3.
Steps for the ultrasonic vibration-assisted hot glass embossing process [2].

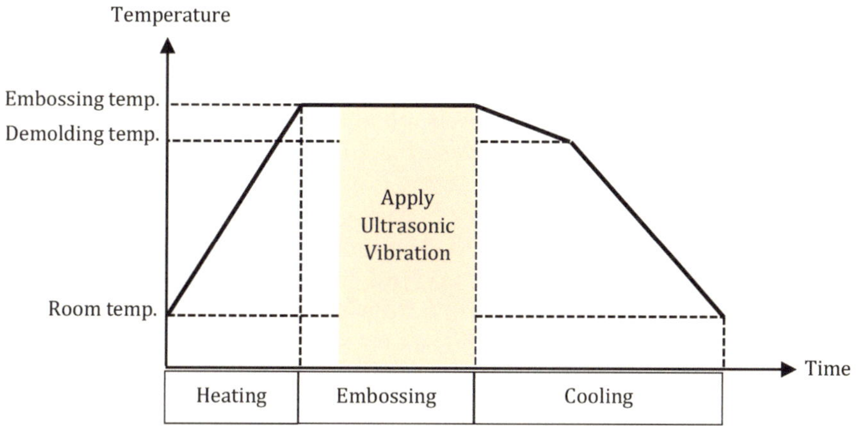

Figure 4.
Temperature setting for the ultrasonic vibration-assisted glass hot embossing process.

1.2 Ultrasonic vibration-assisted hot glass embossing process

Ultrasonic vibration technology has been widely applied in various industrial processes, such as machining, welding, and forming. Recently, this technology has been also utilized for processes working at high temperature like hot upsetting and hot embossing. The steps of an ultrasonic hot glass embossing process are like those of the traditional one, except the embossing stage. During the embossing stage, an ultrasonic source is located on the top of the mold to generate high-frequency vibrations (**Figure 3**). The high energy of ultrasonic vibration rapidly increases the temperature at the contact area between the glass and the mold. Because of this principle, the temperature distribution between the traditional process and the ultrasonic process is different. In the conventional process, the temperature distribution inside the glass during the embossing stage is identical, whereas in the ultrasonic process the localized heat-affected zones are concentrated on the contact area of the mold and the glass. Therefore, this ultrasonic embossing method is not an isothermal process. Temperature setting of this process is shown in **Figure 4**.

2. Components for ultrasonic vibration-assisted hot glass embossing process

In general, an ultrasonic vibration-assisted hot glass embossing process has three main components: heating furnace, compression tester, and ultrasonic vibration device. The role of heating furnace is to heat the glass and the mold to the embossing temperature. Vacuum environment is usually remained within the heating furnace to prevent the heat loss. The ultrasonic vibration device is usually compiled with one of the molds, which transfers ultrasonic vibration to the mold directly. Both the heating furnace and the ultrasonic vibration are attached to the compression tester, as shown in **Figure 5**. The role of compression tester is to control the embossing load after receiving feedback signals from the load cell.

2.1 The heating furnace

A cross-sectional diagram of a heating furnace is shown in **Figure 6**. The heating furnace is integrated by a quartz tube. Because the heating furnace is fixed to the wall, the lower die is controlled by the compression tester to move up and down, so that it can emboss the glass inside the chamber and de-mold the product. Some infrared heaters are distributed around the quartz tube. Energy from the infrared light penetrates through the quartz to heat the molds and the specimen inside.

During the embossing stage, temperature is very high. To protect the load cell, which is located inside the lower die, from damage under high temperature, a cooling system is set up. Besides that, the load cell is also working in the vacuum environment. This condition helps the load cell detect external forces precisely. Although vacuum environment is useful for the load cell, it is useless for the infrared heaters. Therefore, the infrared heaters are placed outside the vacuum chamber to increase their lifetimes.

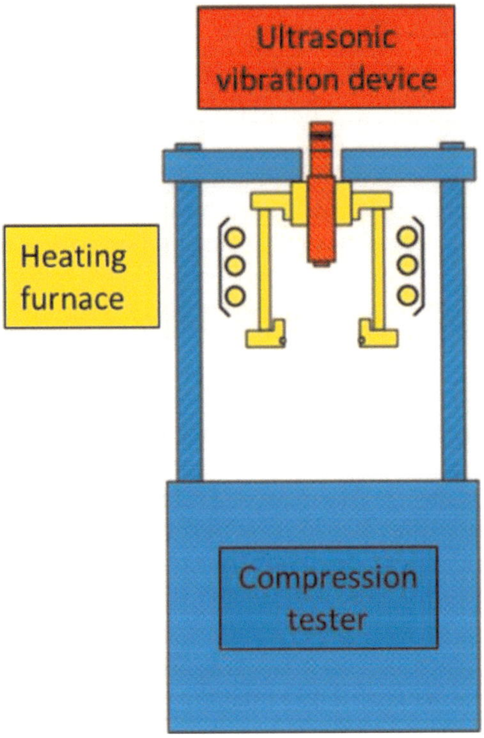

Figure 5.
Schematic of apparatus design [2].

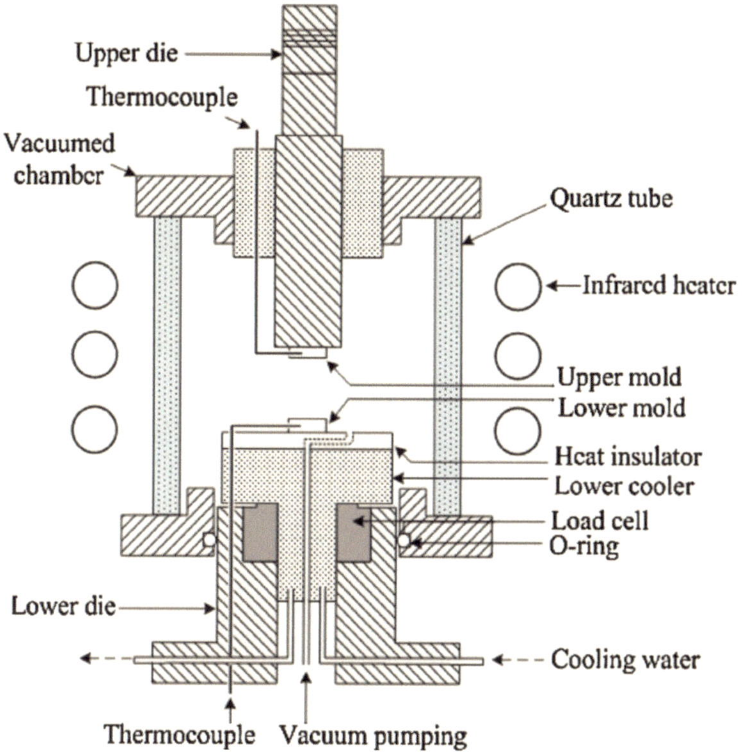

Figure 6.
Cross-sectional diagram of a heating furnace [3].

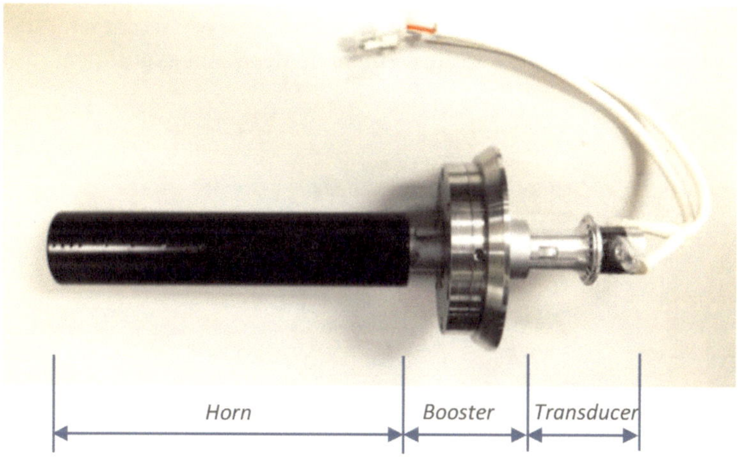

Figure 7.
An ultrasonic vibration device.

2.2 Ultrasonic vibration device

As shown in **Figure 7**, an ultrasonic vibration device consists of a piezoelectric transducer, a booster, and a horn. The vibration is generated from the transducer by inputting an electrical signal through a frequency generator. Resonance phenomenon is usually adopted in ultrasonic vibration devices and then harmonized with the frequency of electrical signals. The ultrasonic vibrating device is designed to work properly at a constant frequency. For thermal protection, a horn cooler is mounted outside the ultrasonic horn. O-rings placed between the ultrasonic horn and the horn cooler to form a water seal do not significantly affect the ability of the ultrasonic device to vibrate.

Since the material properties of the horn would change in elevated temperature, its resonant frequency is shifted, and a mismatch with the frequency generator occurs. Hence, the ultrasonic device must be modified to ensure that it can operate correctly at high temperature. By simplifying theoretical equations, the speed of a wave traveling along a one-dimensional medium is described by

$$c_L = \sqrt{\frac{E}{\rho}} \tag{1}$$

where E and ρ are Young's modulus and density, respectively. The wavelength is

$$\lambda = \frac{c_L}{f} = \frac{\sqrt{E/\rho}}{f} \tag{2}$$

where f is the resonance frequency of the ultrasonic vibration device. In longitudinal vibration mode, multiples of $(\lambda/2)$ can be used as reference for the design of the device length.

As the temperature of a device whose geometry is fixed rises, its resonant frequency falls due to the decrease in Young's modulus [4], so such frequency will shift beyond the tracking range of the frequency generator. To increase the resonant frequency of the device at high temperature, its length must be reduced. With theoretical perspective, reducing the device length could increase the resonant frequency in longitudinal vibration mode. It means that the length reduction could compensate the frequency decrease caused by the increase in temperature. This trend has been verified by both finite element analysis and experiments as shown in **Figure 8**.

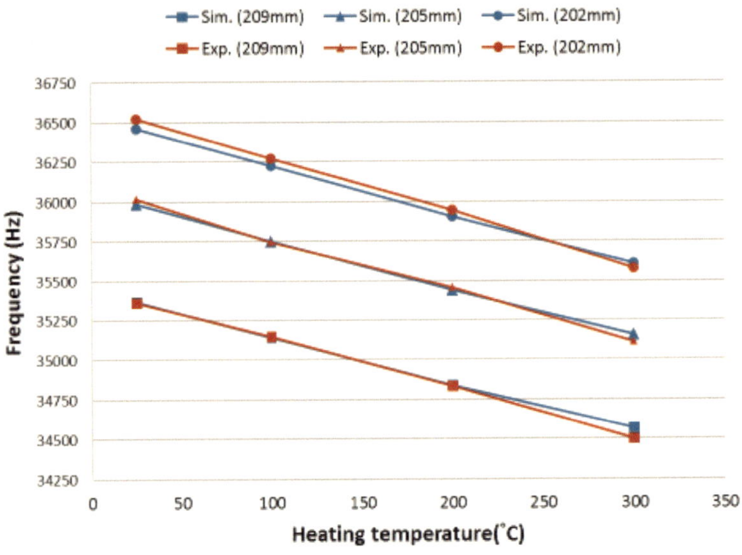

Figure 8.
Resonant frequency of the ultrasonic vibration device with different horns and heating temperatures at 25°C [5].

3. Effects of ultrasonic vibration on hot glass embossing process

3.1 Glass behavior under the application of ultrasonic vibration

In a dynamic experiment, if a sinusoidal strain with angular frequency ω and amplitude ε_0 is applied into a viscoelastic solid

$$\boldsymbol{\varepsilon} = \boldsymbol{\varepsilon}_0\, \boldsymbol{sin}\,(\boldsymbol{\omega t}) \tag{3}$$

the resulting stress would be also sinusoidal with the same frequency which is lagging with a phase angle δ (**Figure 9**):

$$\boldsymbol{\sigma} = \boldsymbol{\sigma}_0\, \boldsymbol{sin}\,(\boldsymbol{\omega t} + \boldsymbol{\delta}) \tag{4}$$

Using complex notation

$$\boldsymbol{\varepsilon}^* = \boldsymbol{\varepsilon}_0\, \boldsymbol{exp}\,[\boldsymbol{i}(\boldsymbol{\omega t})]; \boldsymbol{\sigma}^* = \boldsymbol{\sigma}_0\, \boldsymbol{exp}\,[\boldsymbol{i}(\boldsymbol{\omega t} + \boldsymbol{\delta})] \tag{5}$$

the complex modulus G^* is then defined by the relation

$$\mathbf{G}^* = \frac{\boldsymbol{\sigma}^*}{\boldsymbol{\varepsilon}^*} = \frac{\boldsymbol{\sigma}_0}{\boldsymbol{\varepsilon}_0}\, \boldsymbol{exp}\,(\boldsymbol{i} \cdot \boldsymbol{\delta}) = \frac{\boldsymbol{\sigma}_0}{\boldsymbol{\varepsilon}_0}[\boldsymbol{cos}\,\boldsymbol{\delta} + \boldsymbol{i} \cdot \boldsymbol{sin}\,\boldsymbol{\delta}] = (\mathbf{G}' + \boldsymbol{i} \cdot \mathbf{G}'') \tag{6}$$

The first term on the right-hand side of Eq. (6) is in phase with the strain and is the real part of the complex modulus, often called the storage modulus:

$$\mathbf{G}' = \frac{\boldsymbol{\sigma}_0}{\boldsymbol{\varepsilon}_0}\, \boldsymbol{cos}\,\boldsymbol{\delta} \tag{7}$$

The second term of Eq. (6) represents the imaginary part of the complex modulus, often called loss modulus:

$$\mathbf{G}'' = \frac{\boldsymbol{\sigma}_0}{\boldsymbol{\varepsilon}_0}\, \boldsymbol{sin}\,\boldsymbol{\delta} \tag{8}$$

The ratio $G''/G' = tan\delta$, so-called loss factor, is widely used as a measure of the damping capacity of viscoelastic materials.

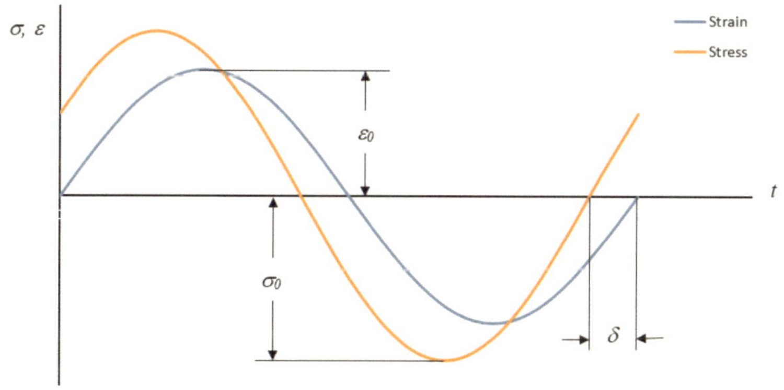

Figure 9.
Oscillating strain ε, stress σ, and phase lag δ.

3.2 Viscoelastic dissipation during the embossing stage of ultrasonic hot glass embossing process

During the embossing stage of the ultrasonic hot glass embossing process, besides embossing load (force or displacement input), high-frequency longitudinal waves are applied to the mold; thus, using this kind of load is similar to applying an oscillating load to the glass material. As propagating through the glass material, the energy of ultrasonic vibration is absorbed and converted into other kinds of energy. Since glass is an amorphous material, energy of ultrasonic vibration should be mostly converted into heat. The amount of heat will cause the temperature rise of both glass and molds, which should be the explanation for experimental observations, including the reduction of embossing force and the improvement of the micro-replication of the glass. In order to model the glass behavior during the

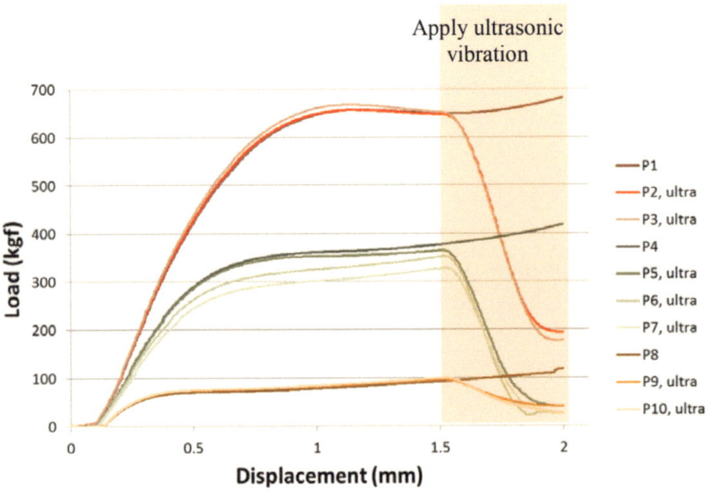

Figure 10.
Force-displacement results for the flat hot embossing experiments [7].

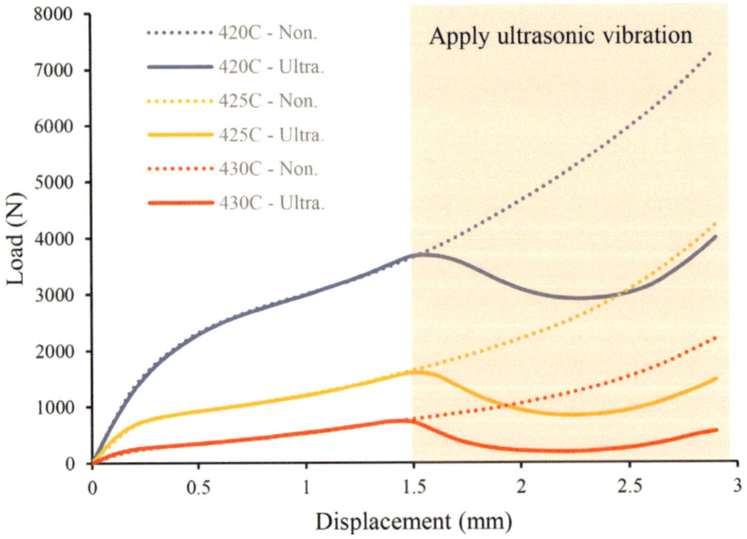

Figure 11.
Reduction of embossing load with different embossing temperatures [8].

embossing stage with the assistance of ultrasound, it must determine the amount of heat which is the conversion of energy of ultrasound absorbed by the glass.

In the previous section, we introduced the loss modulus G" which represents the energy dissipated by the material. The significance of G″ is made apparent by calculating the energy absorbed by the specimen. If dissipated energy is converted to heat completely, then the heat generation rate which would be inputted into the finite element simulations is [6]

$$\dot{q} = f \cdot \pi \cdot G'' \cdot \varepsilon_0^2 \tag{9}$$

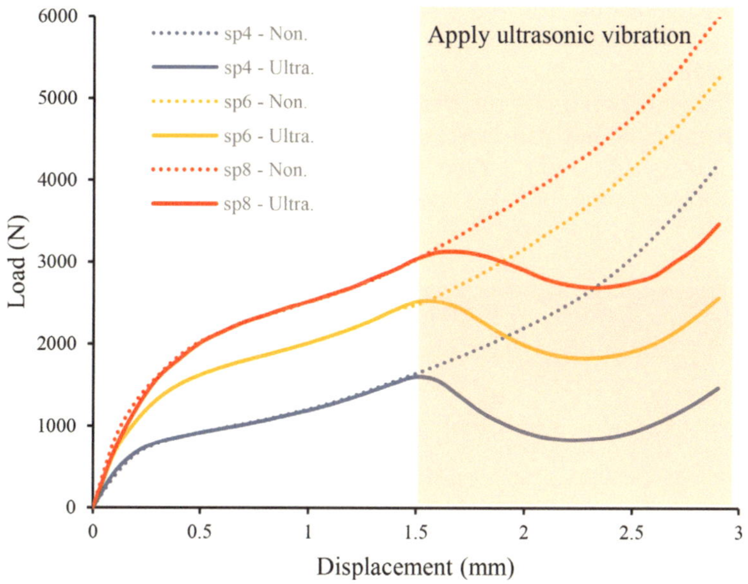

Figure 12.
Embossing load at initial temperature of 425°C and at different speeds [8].

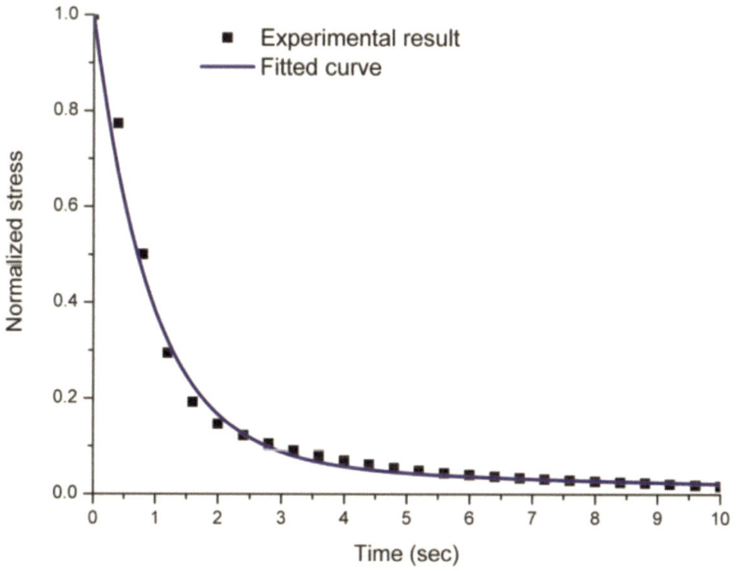

Figure 13.
Experimental result of stress relaxation at 556°C (Tg + 50°C) and the fitted curve [9].

3.3 Effect of ultrasonic vibration on reducing embossing load

During the embossing stage, the mold is controlled by force or by displacement to emboss the glass. In general, at a higher molding temperature, the force required to emboss the glass is lower. As ultrasonic vibration is applied, the forces dropped rapidly, while the displacement continues to increase (**Figure 10**). However, this force reduction could not remain to the end of the embossing stage. After reducing to a finite value, the force increases again (**Figure 11**). This interesting effect was not only verified experimentally with different initial molding temperatures but also with different embossing speeds (**Figure 12**). This phenomenon could be explained by the heating effect of ultrasonic vibration. The high energy of ultrasonic vibration was mainly converted to heat, causing the temperature of the glass specimen to rise.

The force reduction under the effect of ultrasonic vibration is not only a safe solution for the mold but also decreases the molding time of the whole process. The stage which wastes most of the time is the cooling stage. In this stage, the glass needs quite a lot of time for stress relaxation and structural relaxation. The larger

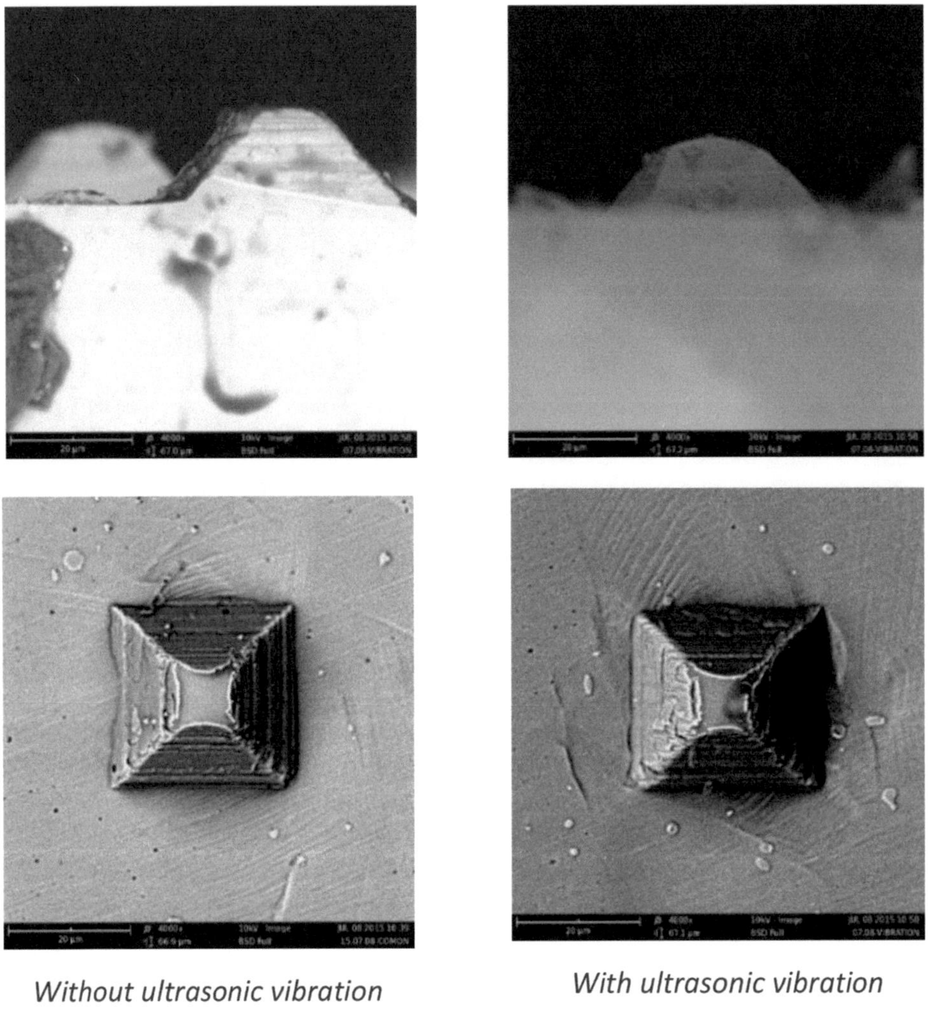

Figure 14.
Scanning electron microscope (SEM) of pyramid structures [10].

the value of embossing load at the end of the embossing stage, the longer the required cooling time (**Figure 13**). Therefore, with the significant reduction force as applying ultrasonic vibration, the productivity of hot glass embossing process could be improved.

3.4 Effect of ultrasonic vibration on improving glass formability

As mentioned above, hot glass embossing is a novel process to produce microstructures on glass substrate. Although the fabrication of microstructure should be performed by conventional process, the accuracy of the final shape of products is hard to achieve due to the surface defect or the adhesion between the glass and the mold as glass is filling into the micro cavities. These disadvantages especially appear

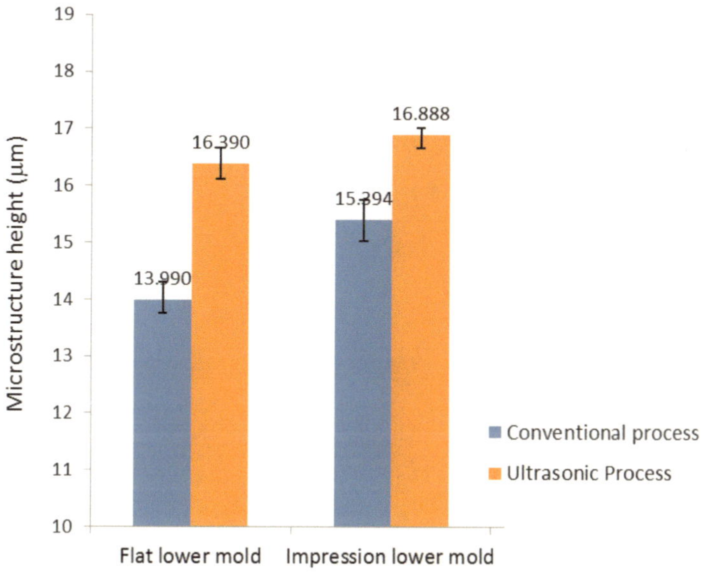

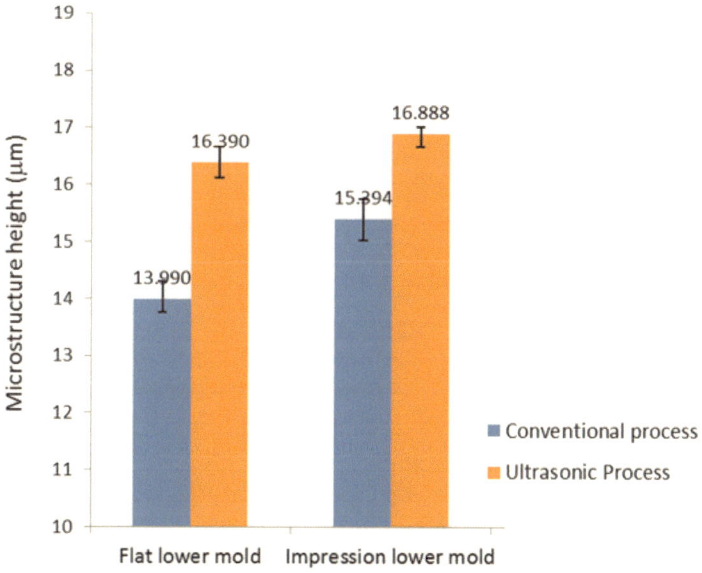

Figure 15.
Comparison of the final height of microstructure between conventional process and ultrasonic process [11].

with three-dimensional microstructures such as micro-grooves, micro-pyramids, micro-prisms, and micro-lenses which are increasingly needed in optical, optoelectronic, and biomedical industries. These difficulties could be resolved with the effect of ultrasonic vibration. **Figure 14** shows the experimental data results which illustrate that applying ultrasonic vibration to hot embossing process can increase the filling ability of glass material significantly (up to 17%). Under the effect of ultrasonic vibration, the glass formability would even improve better than the case of using compression mold, which is usually a solution to help the glass fill more into the microwave (**Figure 15**). Similarly, ultrasonic vibration could also increase

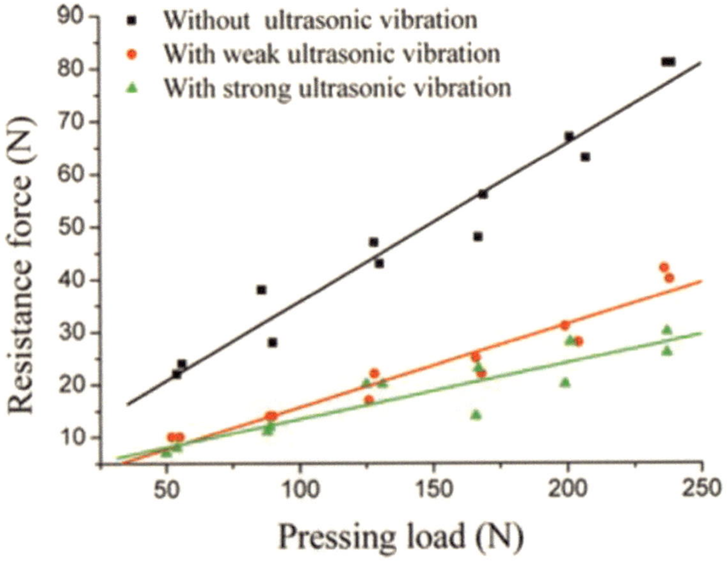

(a)

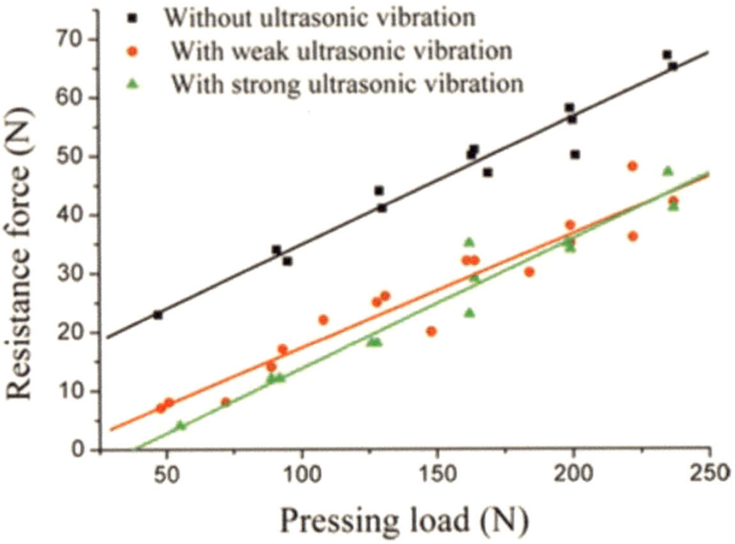

(b)

Figure 16.
Resistance force decrease by ultrasonic vibration with smooth end surface (a) and rough end surface (b) glass [14].

the embossing speed. Experiments show that the amount of glass filled into the micro cavities at high speed during the ultrasonic process was even more than that at lower speed during conventional process [12, 13]. Similar to the above discussion, these phenomena could be explained by the heat generation when glass absorbed the energy from ultrasonic vibration.

3.5 Effect of ultrasonic vibration on decreasing friction force

Glass pressing experiments with the assistance of ultrasonic vibration at room temperature and at elevated temperature have been performed to show effect of ultrasonic vibration on friction force [14]. Some hot embossing experiments were first performed at room temperature. Since ultrasonic vibration is applied to the glass as a sinusoidal displacement, the time that the glass contacts to the mold would be much shorter than that without ultrasonic vibration. Another finding was that the contacting time could be decreased more as increasing the amplitude of ultrasonic vibration. Those findings should be considered as the reasons for the friction reduction during the hot embossing process assisted by ultrasonic vibration [15]. **Figure 16** shows the results from the above experiments. Two kinds of glass with different surface qualities were used as specimens for both conventional and ultrasonic experiments. Although the resistance force grew linearly with the increase of pressing load, the resistance force in case of ultrasonic experiment was much lower than that in case of conventional case. Further, the reduction of resistance force was also more with better quality of glass surface (60% with smooth surface compared to 50% with rough surface in average). As the resistance force is smaller, the life of the mold should be prolonged [16].

4. Finite element analysis of ultrasonic vibration-assisted hot glass embossing process

Glass molding, also hot embossing, is a replicative process that allows the production of high-precision optical components from glass without grinding and polishing. Many researchers have been studying the glass molding process using finite element analysis. However, very few studies have focused on the hot glass embossing process assisted by ultrasonic vibration. Since the only difference between the conventional process and the ultrasonic vibration-assisted process is in the embossing stage, the glass model for the embossing stage should be created. This model could not only describe the glass behavior under embossing force but also express the effect of ultrasonic vibration. Standard linear solid (SLS) model, one kind of viscoelastic models, which combines a Maxwell model and a spring in series, was proposed for the glass deformation behavior during the embossing stage [8] (as shown in **Figure 17**). Substituting complex strain and complex stress from Eq. (5) into constitutive equation [8]:

$$\sigma = \left[\frac{E_0\eta_1}{E_1} - \frac{v}{L}\frac{(E_0 + E_1)}{\left(\frac{E_1}{\eta_1} - 1\right)}\right] exp\left(-\frac{E_1}{\eta_1}\varepsilon\right) + \frac{v}{L}\frac{(E_0 + E_1)}{\left(\frac{E_1}{\eta_1} - 1\right)} exp\left(-\varepsilon\right) + E_0\varepsilon - \frac{E_0\eta_1}{E_1} \tag{10}$$

where E_0, E_1, and η_1 are constants, determined by fitting experimental data; v is the embossing speed; and L is the initial height of glass sample. Storage and loss moduli can be calculated as

$$G' = E_1\left[\frac{(\omega\eta_1{}^2)}{E_1^2 + (\omega\eta_1)^2}\right]; G'' = E_1^2\left[\frac{\omega\eta_1}{E_1^2 + (\omega\eta_1)^2}\right] \tag{11}$$

where ω is the angular frequency of ultrasonic vibration. Substituting G'' in Eq. (11) into Eq. (9), the amount of heat created by ultrasonic vibration could be determined.

Another viscoelastic model, which has been proposed for the glass deformation behavior during the hot embossing stage, is the Generalized Maxwell model, as shown in **Figure 18** [14]. Basically, Generalized Maxwell model is similar to SLS model, except having more the elements of Maxwell model.

The time-dependent response is characterized by the deviatoric terms as

$$\sigma = \int_0^t 2G(t-\tau)\frac{de}{d\tau}d\tau \tag{12}$$

The above integral is evaluated for current time t on the basis of past time τ. G(t − τ) is not a constant value, but it is represented by a Prony series, as described by

$$G(t-\tau) = G_0\left[\sigma_\infty + \sum_{i=1}^{n_G} \alpha_i \, exp\left(-\frac{t}{\tau_i}\right)\right] \tag{13}$$

where τ_i is the relaxation time, α_i the weight factor, n_G the number of Generalized Maxwell model units, and G_0 the initial modulus. The dynamic viscosity η^* can be calculated correspondingly, as demonstrated by

$$\eta^* = \sum_{i=1}^{n_G} G_i \frac{\eta_i}{1+\omega_i^2\tau_i^2} \tag{14}$$

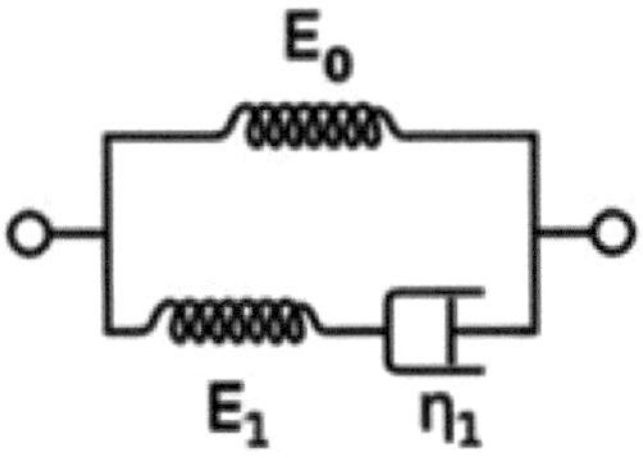

Figure 17.
Standard linear solid model.

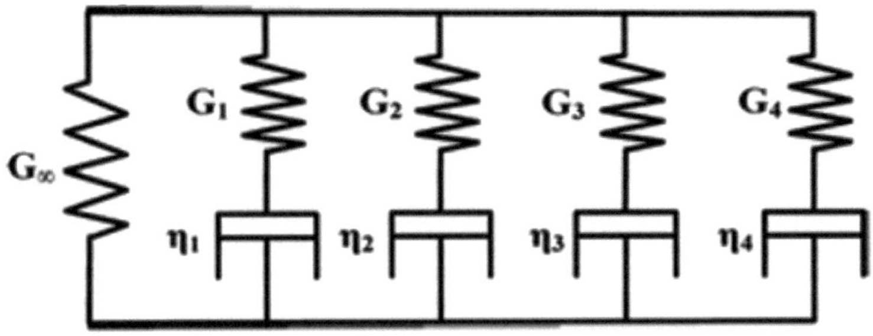

Figure 18.
Generalized Maxwell model.

As an alternating stress is applied to the material specimen, it is similar to an elastic solid which its elastic properties are exhibited, and the dynamic viscosity decreases significantly. The above reason shows that ultrasonic vibration could be assumed to improve the forming process [15].

After inputting the above models to simulation, simulated results could be used to verify the value of proposed models. As shown in **Figures 19** and **20**, the agreement between simulated results and experimental data proved that finite element analysis would have an important role in analyzing and predicting the effect of

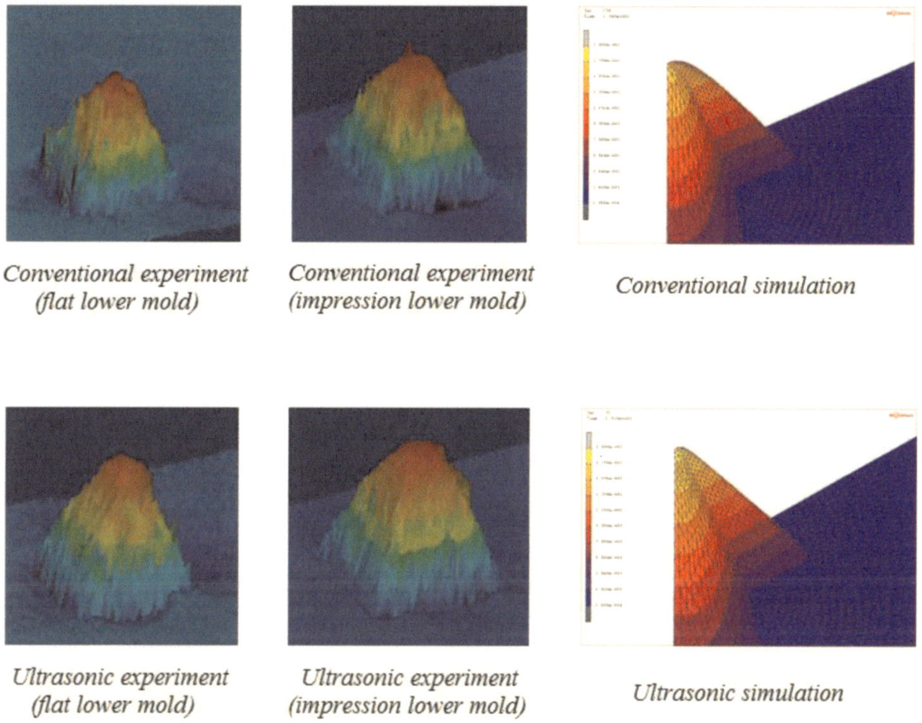

Figure 19.
3D profile of pyramid microstructures after hot embossing process [10].

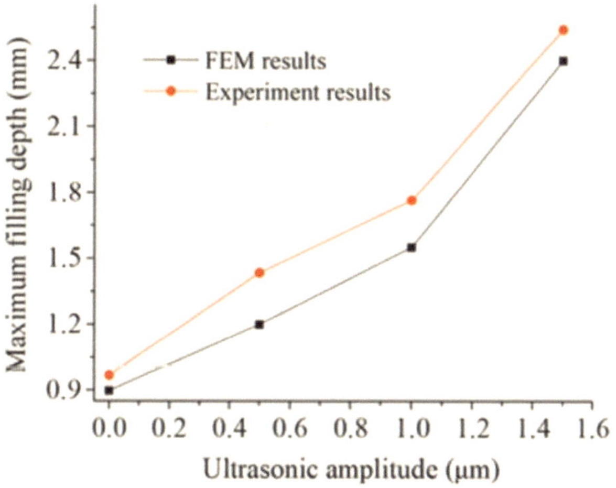

Figure 20.
Plot of glass maximum filling depth against amplitude [14].

process factors, such as temperature, speed, amplitude, frequency, etc. in improving the quality of the final glass products.

5. Conclusion

Ultrasonic vibration technology has been utilized in hot glass embossing process. Under the effect of ultrasonic vibration, glass formability could be improved significantly. Moreover, productivity of the whole process also increases, and life time of molds could be expanded. Finite element analysis has become an effective tool for analyzing and predicting so that the quality of the final glass products would be achieved.

Acknowledgements

I would like to express my sincere gratitude to Prof. Chinghua Hung and all members in Precision Engineering and Simulation Laboratory, Department of Mechanical Engineering, National Chiao Tung University (Taiwan), for their support in all the time of writing this manuscript.

My sincere thanks also go to my husband, my son, and all members of my family for their encouragement so that I could finish this manuscript.

Conflict of interest

The author declares that there are no affiliations with or involvement in any organization or entity with any financial interest or non-financial interest in the subject matter or materials discussed in this manuscript.

Author details

Lanphuong Nguyen
Hanoi University of Science and Technology (HUST), Hanoi, Vietnam

*Address all correspondence to: phuong.nguyenlan@hust.edu.vn

References

[1] Worgull M. Hot Embossing: Theory and Technology of Microreplicaiton. Oxford: William Andrew; 2009

[2] Tsai Y-P. Ultrasonic Vibration-Assisted Optical Glass Hot Embossing Process. [dissertation] Hsinchu: National Chiao Tung University; 2013

[3] Hung J-C, Tsai Y-P, Hung C. Development of a new apparatus for ultrasonic vibration-assisted glass hot embossing process. Precision Engineering. 2013;**37**(1):222-227

[4] Gardner L, Insausti A, Ng KT, Ashraf M. Elevated temperature material properties of stainless steel alloys. Journal of Constructional Steel Research. 2010;**66**(5):634-647

[5] Nguyen LP, Tsai Y-P, Hung J-C, Hsieh Y-C, Hung C. Finite element analysis of an ultrasonic vibration device at high temperatures. Journal of the Chinese Society of Mechanical Engineers. 2016;**37**(3):193-200

[6] Zarzycki J. Glasses and the Vitreous State. New York: Cambridge; 1991

[7] Tsai Y-P, Hung J-C, Yin L-C, Hung C. Ultrasonic vibration-assisted optical glass hot embossing process. The International Journal of Advanced Manufacturing Technology. 2012; **60**(9-12):1207-1213

[8] Nguyen LP, Hao K-C, Su Y-H, Hung C. Modeling the embossing stage of the ultrasonic-vibration-assisted hot glass embossing process. International Journal of Applied Glass Science. 2015; **6**(2):172-181

[9] Tsai Y-C. Finite Element Analysis on the Optical Glass Molding Process. [dissertation] Hsinchu: National Chiao Tung University; 2010

[10] Nguyen L. Finite Element Analysis of the Ultrasonic Vibration-Assisted Hot Glass Embossing Process. [dissertation] Hsinchu: National Chiao Tung University; 2015

[11] Nguyen L, Wu M-H, Hung C. Effect of ultrasonic vibration on microstructure hot glass embossing process. Journal of Science and Technology. 2017;**122**:17-21

[12] Nguyen L, Wu M-H, Hung C. Effect of Ultrasonic Vibration on Increasing Embossing Speed during Hot Glass Embossing Process. In: Proceedings of the First International Conference on Materials, Machines and Methods for Sustainable Development (MMMS2018). Da Nang; 18-19 May 2018;**1**:98-105

[13] Nguyen L, Wu M-H, Hung C. Effect of ultrasonic vibration on increasing embossing speed during hot glass embossing process. Applied Mechanics and Materials. 2019;**889**:71-79

[14] Zhou T, Xie J, Yan J, Tsunemoto K, Wang X. Improvement of glass formability in ultrasonic vibration assisted molding process. International Journal of Precision and Manufacturing. 2017;**18**(1):57-62

[15] Xie J, Zhou T, Liua Y, Kuriyagawa T, Wang X. Mechanism study on microgroove forming by ultrasonic vibration assisted hot pressing. Precision Engineering. 2016;**46**:270-277

[16] JinyaoChen Y, Chen H, Lia S-Y, Jowc J. Physical and chemical effects of ultrasound vibration on polymer melt in extrusion. Ultrasonics Sonochemistry. 2010;**17**(1):66-71

Chapter 3

mDFA Detects Abnormality: From Heartbeat to Material Vibration

Toru Yazawa and Shinji Omata

Abstract

Modified detrended fluctuation analysis (mDFA) is a novel method to check abnormality of heartbeat which is developed recently by the author. mDFA can characterize any oscillation such as heartbeat by the scaling exponent (scaling index, SI). Healthy heartbeat shows SI = 1. Dying heart's SI sifts toward 0.5. Ischemic sick heart experimentally showed an SI way over 1.0 approaching 1.5. Random vibration, such as FM-radio noise and idling car-engine, shows SI = 0.5. Quietly running motor generates an SI almost equal to zero. Using mDFA, it is possible to check potential risk based on SI values. This chapter shows empirical results quantifying various signals from heartbeat to material vibration.

Keywords: mDFA, heartbeat rhythm, material vibration, quantifying risks, scaling exponent

1. Introduction

System failure—ischemic disease of the cardiac system (CS), highway bridge crash, earthquake, space rocket launch failure, and so forth—leads to catastrophic consequences. Watching abnormality by testing toughness is important for preventing disasters. We want to detect early warning sign. But how? We have an idea. The method is modified detrended fluctuation analysis (mDFA).

In this chapter, we show examples of mDFA computation. First, we show mDFA results on the CS, then results on the nonliving, material system.

The CS is composed of the heart and the brain, that is, a pump and a controller. Evolutionally, all creatures have evolved from a common ancestor. If the CS was innovated long ago, "hearted" animals follow fundamental laws of physics (and chemistry and biology). In fact, it is known that all "hearted" animals carry the same DNA for making the heart: development of vertebrate heart is controlled by a common genetic code (a DNA sequence called Nkx, a homeobox, and a cardiac-specific homeobox). However, not only vertebrate heart but also invertebrate heart is controlled by the same genetic code: development of insect heart is regulated by "tinman DNA," which is Nkx family gene. Surprisingly, Hydra, a simple animal, that exhibits "pumping movement," has Nkx-like gene (see [1]). Therefore, the pumping-heart design is evolved since Hydra [2]. If we find a primordial basic rule in a simple creature, it is applicable to humans.

In this chapter, we first show experimental results on crustacean animals (crabs and lobsters), on which we have long worked [3]. In the second place, we present human heartbeat analysis, and lastly mechanical vibration analysis. Throughout the study, we use an analytical method, mDFA, which is recently developed by our group [1]. We explain mDFA in the later section.

2. Physiology of animal heart

Circulation failure is absolutely a life-threatening event. Unpredictable cessation of blood flow is the worst-case scenario. For studying the problems, we need physiology, and it is necessary to record heartbeat data from freely moving animals.

Figure 1 shows how to record electrocardiogram (EKG) from invertebrate animals. Two permanently implanted metallic electrodes (+ and −), touching the surface of the heart with extreme caution, were used to record EKG. We used DAM50 C-R coupled AC amplifier (World Precision Instruments, USA) and Power Lab (ADIntrument, Australia) for digital EKG data sampling at 1 kHz.

Nervous regulation of the crustacean heart is well documented by great scholars [1]: Carlson (1904), Alexandrowicz (1932), Maynard (1961), and so forth. The heart receives two kinds of nerve fibers. One is acceleratory nerve (CA) and the other is inhibitory nerve (CI) [3]. The nerves are always active and discharge frequency is ever-changing. Moreover, the brain releases slowly functioning cardio-active substances (peptide) via the nonneuronal hormonal method [6]. As a result, the heart never beats at a steady pace. Heart rate exhibits a dynamic change all through life. Heartbeat interval time is never stable, never regular, and is always fluctuating.

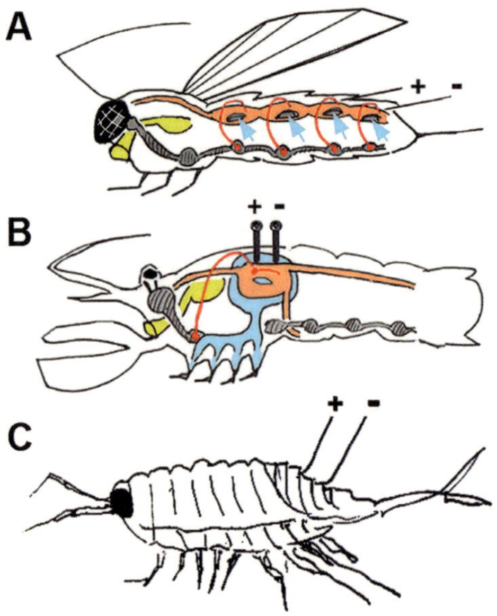

Figure 1.
Diagrammatic representation for EKG recording from insect (A), lobster (B), and isopod crustacean Ligia (C).

3. Analysis of heartbeat

Mechanisms of cessation of heartbeat could be studied by mathematical methods. We believed that the method might be the frequency analysis because heartbeat is a cyclic behavior. The other candidate method is heartbeat-interval time series analysis. Whichever, we need natural data, EKG. We prepared two specimens, intact and isolated hearts. We tested two different analytical methods. One is the power spectral density analysis (PSD) and the other is mDFA [5, 6]. As a result, we found that PSD did not well distinguish the difference between the two heartbeats [5, 6]. In contrast, mDFA was powerful and quantitative to represent the inherent state of the two hearts [4].

PSD is well known worldwide. People who use PSD implicitly suppose that a complex-look signal is a sum of cosine waves at various frequencies such as 10, 20, 30, 40, and 50 Hz. Real world data, such as heartbeat and material-vibration signal, carry hidden information that PSD might not capture.

With this consideration, the mDFA program was invented by a master-student, Tanaka (see [7]), and an author (TY) tested and verified it [1]. It is about 20 years since then.

We repetitively confirmed [1, 4, 5] that freely moving animals' heartbeat exhibits the scaling exponent (SI, scaling index) of around 1 (SI = ~1.0). In turn, isolated heartbeat data exhibit the SI of around 0.5 (SI = ~0.5).

4. Animal heart experiments

At a very early stage of the study, we learned that mDFA well distinguishes between intact and isolated hearts as aforementioned. We got an idea: mDFA can be a helpful tool in pathophysiology, because cardiac disease is one of the major causes of death worldwide.

We began to record long-term EKGs from model animals. The recordings were started from fresh healthy specimens and were kept continued to the end of their life. Sometimes, the recording period length exceeded 2 years, which is extremely long and painstaking (**Figure 1B**, lobster). In turn, it was, at one time, only 2 h (**Figure 1C**, isopod Crustacea).

Figures 2 and **3** show example EKGs at terminal conditions.

Figure 2 shows coconut crab EKG. This specimen was captured in March at a Japanese tropical island, south-west Okinawa. We transfer it to Tokyo in a hand-carried baggage. Long-term EKG was recorded in Tokyo instead of the south, tropical zone. The animal eats apples and dry fish meat and lived longer than we expected. Climate in Tokyo got colder in autumn and the tropical crab ended its life in October: non-air-conditioned environment at natural room temperature. SI values in March were ~1.0 (data not shown). The SI values (around 0.9) continued to September.

In **Figure 2**, one can see that SI values decrease when dying. It is of interest that after the cessation of pumping heartbeat, fibrillation remained (**Figure 2**), which indicates that heart muscles still try to contract.

Many other specimens tested, including crayfish, crabs, insects, and clams, show a SI-decrease-phenomenon when dying (data not shown). We found it typical that when dying, animals show diminished movements and decremental SI-shift toward 0.5.

At the terminal condition, the brain is not likely to regulate the heart any longer, although the heart is still pumping like the isolated heart. We consider that the terminal condition accompanied by a low SI is a state of brain death.

If we look at dying crab specimens, our intuition tells that relevant specimen is likely to pass away soon. We define this as "natural death."

In the meantime, we encountered an unforgettable specimen that died unpredictably (**Figure 3**). At time zero in **Figure 3**, EKG trace looks normal. After checking EKG on PC screen, an author (TY) left Tokyo, setting out on a journey to see a hospitalized family. Two days later, TY returned and looked at PC and discovered that the crab

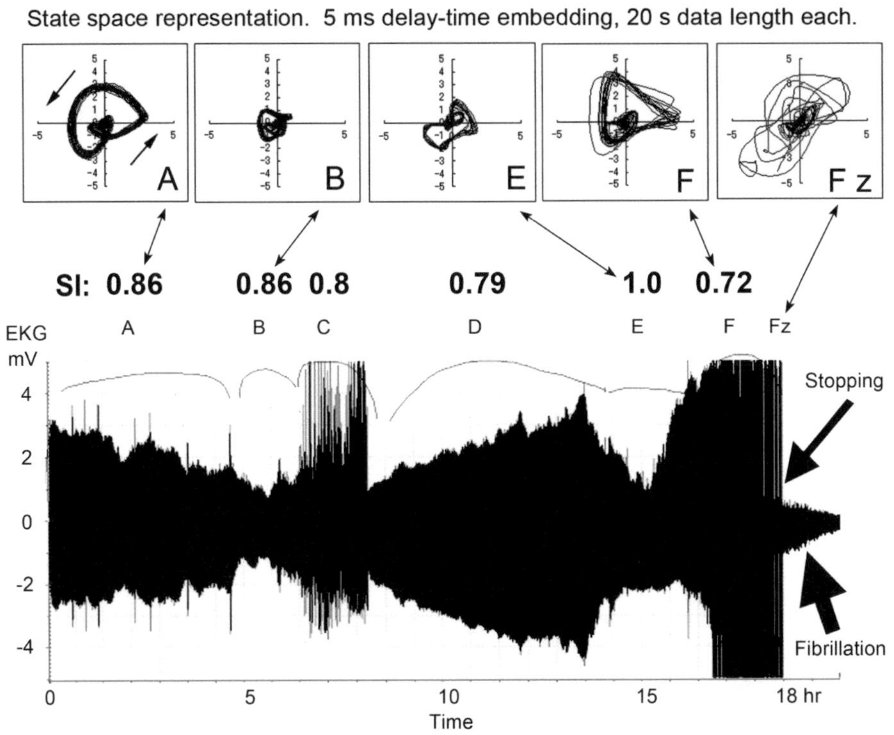

Figure 2.
A long-term EKG from coconut crab (Birgus latro) with state space representation. Recording for 18 h. Inset: SI values for the corresponding period from (A) to (F). Note: fibrillation after beating stops. State space representation of cardiac action potentials shows a normal action potential shape in (A), gradually changing to distorted pattern (B, E, F), and finally becomes erratic and unstable at the end (Fz). Modified from Yazawa [10], Chapter 2.

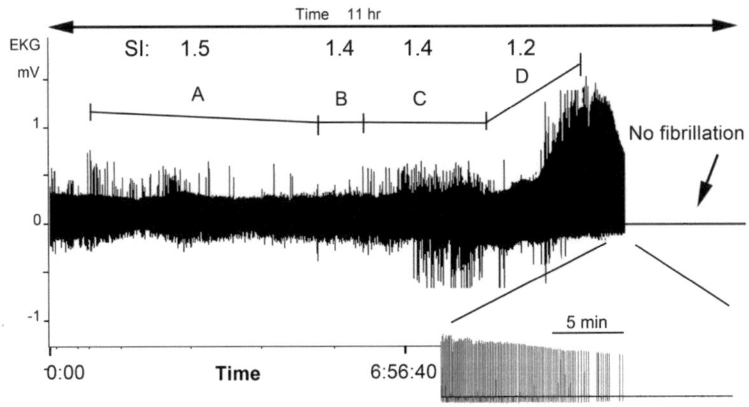

Figure 3.
A long-term EKG from "Mokuzu" crab (Eriocheir japonica). Recording for 11 h. Inset: SI values for the corresponding period from (A) to (D). Note: No fibrillation remained after beating stops: cf. **Figure 2.** *Inset: spike configuration not distorted. Modified from Yazawa [10], Chapter 2.*

heart stopped its beating 11 h after TY left (**Figure 3**). We define this "unpredictable death." Unpredictable death is a rare event among ~1000 specimens. We always check and dissect all specimens' body after the death. In case of **Figure 3** animal, dissection revealed that the myocardium of relevant crab partially got slightly injured by an EKG electrode (see the electrode in **Figure 1B**). It is the fault of researchers. We are very sorry that the innocent specimen suffered from the human-caused heart injury for 2 weeks, although we did not know that. From this rare case, we learned that little injury of myocardium causes sudden death. Only partial damage of the heart is life-threatening. It is comparable to a human health problem, well known ischemic heart disease. We found it crab's "unpredictable death."

Results of **Figure 3** surprised us, the suffering crab's SI values "always" exceeded 1.0 (see **Figure 3**, SI = 1.5, 1.4, 1.4, 1.2). We have never seen before. A hypothesis came into my head: "injured heart has a very high SI." This hypothesis is unproven as far as we know. We have seen many dying specimens that showed reduction of SI-value as shown in **Figure 2**.

The results suggest that "the scaling exponent methods" distinguish damaged heart from unhurt heart. We biologists consider that medical profession should test mDFA on human hearts. We sent a short abstract to a medical congress in Europe, but the application was rejected immediately. We ourselves began to study human hearts.

5. Biophysics

5.1 Quantitative analysis

In 1982, Kobayashi and Musha reported that healthy hearts exhibit 1/f spectrum [8]. Mathematically, 1/f slope is almost equal to SI = 1.0, while not 100% equivalent. This metric analysis based on SI is not fully proven as far as we know. The criterion-based strategy is better than qualitative research for diagnosing the CS.

For the quantitative expression for the CS state, we need computation. However, in the 1980s, we did not have a PC to calculate SI. Poor biologists found difficulty to use a computer that was installed in a building of a top university. It took until 2001 to prove the idea of mDFA by ourselves. The Windows XP machine was introduced in the year 2001. XP-PC helped us calculate mDFA on the second time scale.

What we liked was Kobayashi-Musha's concept that one (SI = 1.0) is "healthy" [8]. There is a fixed baseline for diagnosing the heart system, that is, healthy or not.

In the 1990s, Peng and Goldberger and others demonstrated that healthy hearts exhibit the scaling exponent ONE (1.0) by detrended fluctuation analysis (DFA) [9] [DFA is not mDFA (see below)]. These results add critical evidence to the issue of Kobayashi-Musha's concept.

Moreover, Peng and Goldberger's group reported that sick hearts exhibit a higher SI, which is SI = ~1.2. It sounds like providing evidence that a sick heart was consistently higher in the scaling exponent. But they only suggested. The truth was unclear. At least we were excited about our crab's high-SI discovery (**Figure 3**) because it coincides with it.

Most of the data in [8] and [9] were obtained from in-hospital patients. Peng and Goldberger's group did not extend their detailed experiments to general population as far as we know.

One is a baseline number for the health. This is testable hypothesis. We began to examine EKGs on general population and model animals. Currently we have ~500 individuals' EKG and ~1000 animal data. Some EKGs are collected from long-term follow up. Some subjects have passed away. We take medical record with all data including animal data. We never use website data. Physiological interpretation of data is impossible without medical record based on our own physiological observation.

5.2 Accurate data sampling

EKG signal was captured by two commercially available Ag/AgCl electrodes and a lab-made amplifier that can damp undulatory noise. The amplifier has a short input-time-constant (*tau* = 0.1 or 0.22 s, depending on capacitor used, 0.01 or 0.022 μF). The *tau* value for EKG-machine in hospital-use is set at about 10 s under the international regulation. A large *tau* amplifier makes signal-baseline drift when subjects move. We thus needed to make a small-*tau* amplifier in order to reduce "movement-induced noise and drift" (see **Figure 4**).

Peaks are captured automatically by a lab-made program. R-peaks are the time point at which V_{max} is attained over threshold voltage. Once R-peaks were captured, all peaks were 100% affirmed by eye observation on PC screen after the end of recording. If incorrect peaks are captured, or correct peaks are NOT captured, we

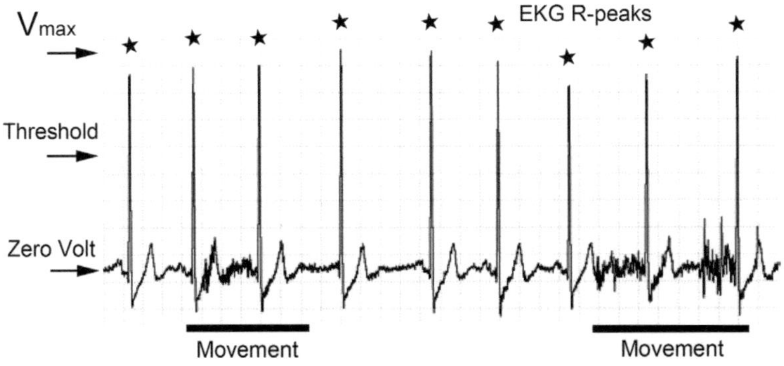

Figure 4.
Accurate peak detection. Human EKG recorded with a lab-made amplifier. A physician diagnosed this heart as a sinus arrhythmia, but not life-threatening; male age: 60s.

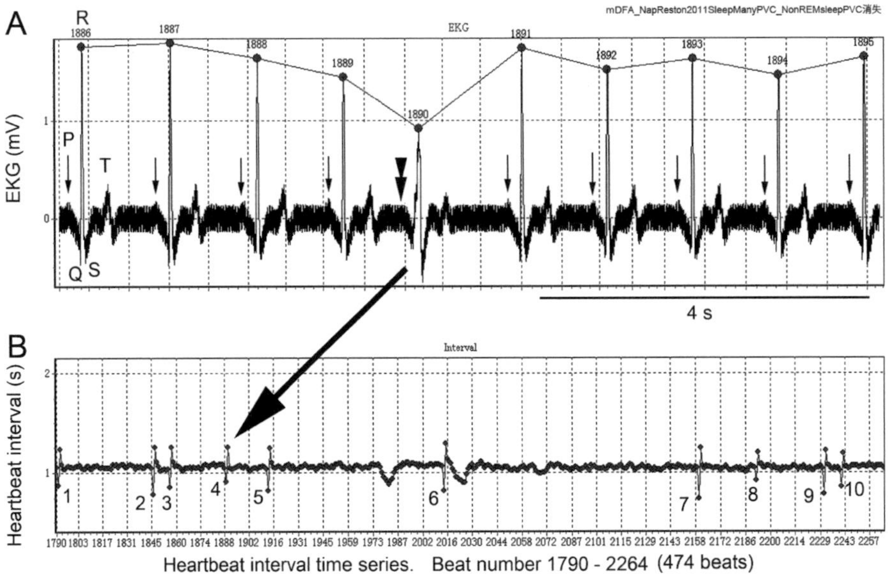

Figure 5.
Accurate data collection. Peak identification from EKG (A) and construction of time series (B). P, Q, R, S, and T peaks are indicated (see A). Small arrows in A point all the P-peaks. Double arrowhead indicates that there is no P-peak within one heartbeat period. Therefore, this is premature ventricular contraction (PCV). Within 474 beat time period, 10 PVCs are visible. **Figure 5** *was recorded 5 years before* **Figure 4** *from the same subject.*

manually made a correction on PC screen. As a result, all peak interval time series become accurate. It is time consuming work but inevitable for accurate interpretation of data. The sampling rate is 1 kHz for the heartbeat (20–40 kHz for material-vibration as shown in below sections).

Since we consider that accurate data sampling before analysis is paramount for later interpretation of results, skipping heartbeats and irregular heartbeats should not be deleted before analyzing, like someone does. However, any artificial spiky noise should not be counted as a pulse.

In summary, accurately recorded EKG without large noise, accurately captured R-peaks (stars in **Figures 4** and dots in **Figure 5A**), and accurate peak-to-peak time interval time series is important for performing accurate mDFA (**Figure 5B**).

5.3 Time series

Figure 6 shows the procedure of mDFA.

5.3.1 First procedure

We first construct an accurate R-R interval time series from EKG recordings, which is [*Xi*] (**Figure 6A**, abscissa axis, the number of heart beat *i*, and vertical axis, rate of beating in beat per min). **Figure 6** shows only 1 - 30 beats among 2000 beats. We use "heart rate" instead of "interval time." If we use R-R-interval time, which is an inverse of "rate," mDFA results are the same. To biologists, using "rate" is intuitively more understandable about physiology of the heart than using "interval." The seventh beat in **Figure 6A** shows an irregular beat.

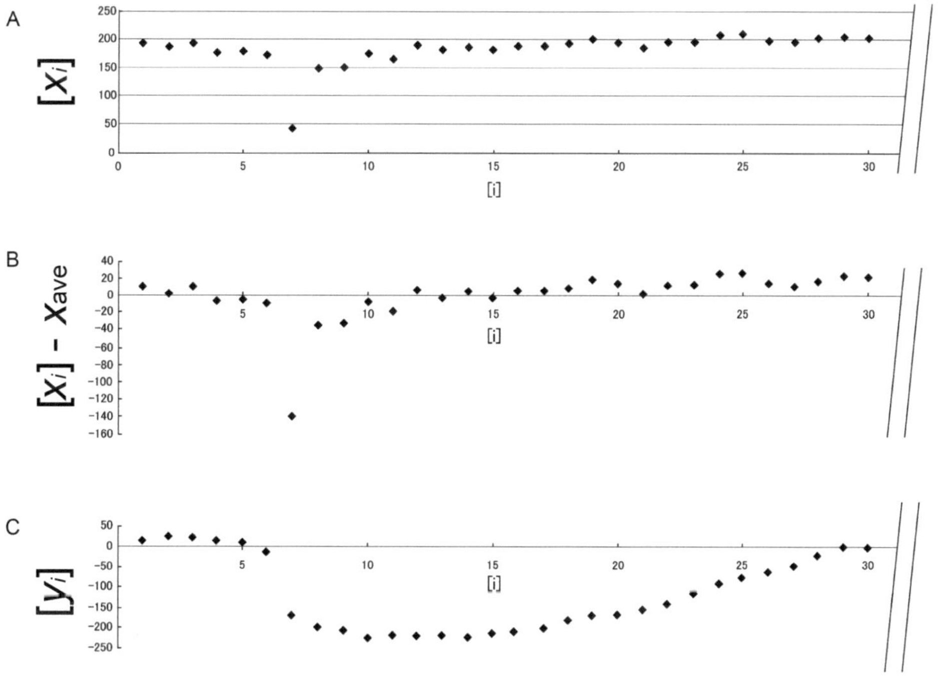

Figure 6.
A diagrammatic explanation of pretreatment of R-R peak data. Measuring a R-R peak interval time Xi, where i = 1, 2, 3, --- 2000 (A), obtaining an average of them, Xave, and thereafter subtracting it from Xi (B), and making additions of each values one by one (C).

For mDFA computation, we use 2000 heartbeat data. Both data shorter or longer than 2000 can be usable but we fixed it 2000 after testing [1]. Long data, such as 1 or 2 h data, does not have significant benefit for interpretation of physiological meaning of results. The reason is simple. The cardiac system (CS) never becomes a stable state. The CS is an ever-changing dynamic system, which is our temporary interpretation and we have had consistent results. Currently, we use 2000 beat data. A 2000 beat time period length is about 30–40 min [1].

5.3.2 Second procedure

Mean value from 2000 data is X_{ave}. By removing X_{ave} from each data (X), one can get a time series of pure fluctuation $[X - X_{ave}]$ (**Figure 6B**).

5.3.3 Third procedure

A computation $\sum_{1}^{2000} x$ makes a random-walk like temporal sequence $[y_i]$ (**Figure 6C**). Important concepts in mDFA are "averaging" and "sigma (summation of data, **Figure 6C**)."

5.4 Trend

Figure 6 demonstrates diagrammatically that the fluctuation property is expressed in connection with the average value. The sequence $[x_i]$ is heart rate time series in beat per min (**Figure 6A**). The sequence $[X_i - X_{ave}]$ expresses pure fluctuation (**Figure 6B**), some larger and some smaller than the average value. One can see that the seventh beat in **Figure 6A** shows a very small value. The seventh beat makes $[y_i]$ trace jump down (see the seventh dot, i = 7, in **Figure 6**). It is catastrophic happening; thus, this event is an arrhythmic heartbeat. This kind of event becomes a matter of life or death if extremely unlucky. In fact, a single event is not only life threatening but also not so happy of course. Therefore, the trait of fluctuation is directly linked to life or death.

In summary, the sequence $[y_i]$ expresses sigma (Σ) of each value. This $[y_i]$ is "trend." This is an explanation about the "pretreatment" of data before conducting mDFA. This $[y_i]$ is the data that mDFA analyzes. Both mDFA and DFA use $[y_i]$ for calculation, but the concept is different between them as shown below. See [1] for details.

5.5 Box size

In **Figure 7A**, 2000 beat long data are broken up into small length data; here it is 10 beat long (see three Boxes in **Figure 7A**). Box-size is freely changeable in program. In **Figure 7**, we show only box-size-10 as an example. We tested smaller box less than10 in box-size. As a result, it is not so useful than we thought. In our program, mDFA's box size ranges from 10 to 1000 [1]. In computing, mDFA automatically changes box-size, starting from box-size 10-beat. Then 11-beat, 12-beat, 13, 14, 15, 16, 17, 18, 19, 20, 21, 22, ..., 30, 31, 32, ..., 40, 41, ..., 50, 51, ..., 60, 61, ..., 70, 71, ..., 80, ..., 90, ..., 100, then 110, 120, and so forth [1].

5.6 Fitting curve

Figure 7B shows a fitting curve. They are linear fitting curves $y_v(1)$, $y_v(2)$, and $y_v(3)$ (**Figure 7B**). This example (**Figure 7B**) is linear fitting, just for the sake of ease. But, in practice, we must use biquadratic fitting [1] (see **Figure 8** caption).

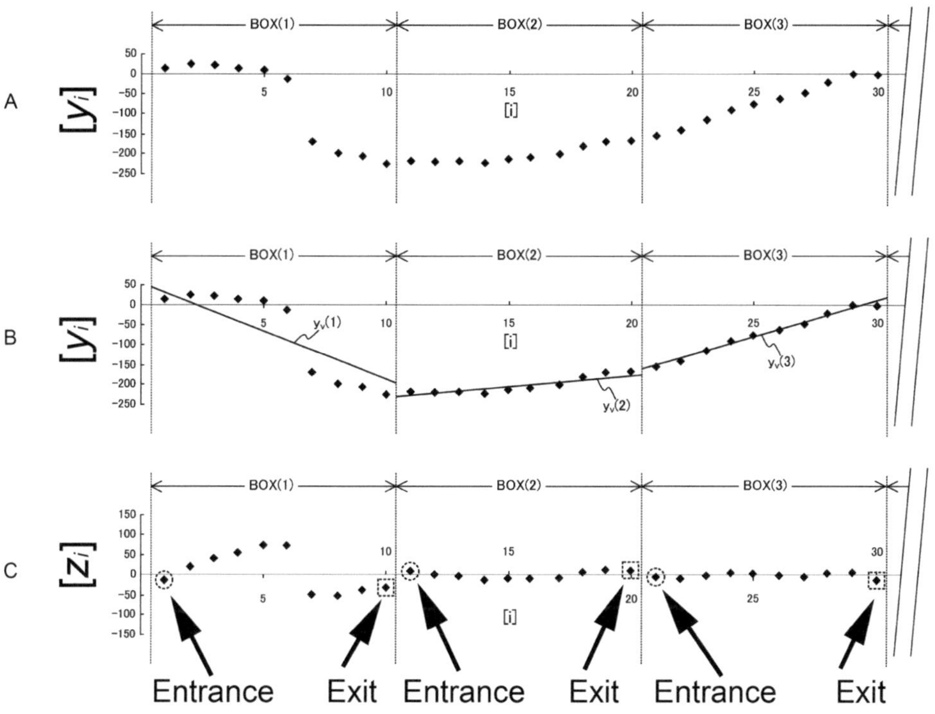

Figure 7.
Diagrammatical representation of the "detrending" procedure. A series shown in ***Figure 6C*** *is broken up into 10-beat-long box (A), drawing a fitting curve in each box (B), and thereafter executing detrending (C). See text.*

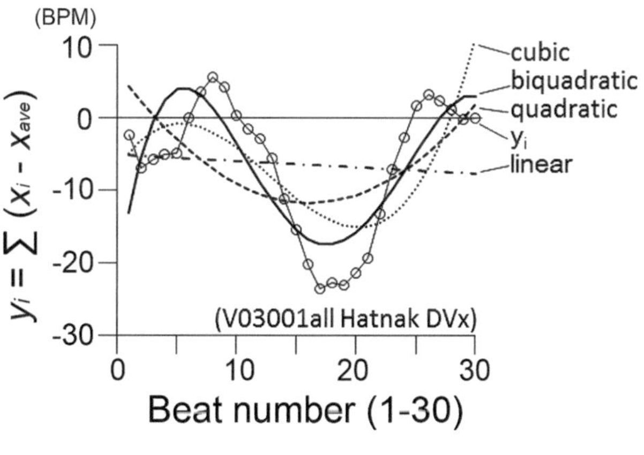

Figure 8.
Several kinds of fitting; real human data. The [y_i] curve shows raw data. We tested fittings, linear, quadratic, cubic, biquadratic, and further. As a result, less than fourth order computation does not return a stable/ unvarying SI value. mDFA uses quadratic fitting.

5.7 Detrending

A fitting y_v curve is given by a PC computation as shown in **Figure 7B**. Next computation is making a detrended curve that is given as $z_i = y_i - y_v$. After this procedure, true fluctuation remains. This is "detrending." The [z_i] sequence is important.

5.8 DFA and mDFA

In **Figure 7C**, arrows point two values: "entrance value" and "exit value" in each box (**Figure 7**). mDFA uses entrance and exit values obtained by detrending procedures.

In turn, Peng's DFA uses all 10 values in each box (see **Figure 9**). But the detrending procedure is a common concept between mDFA and DFA.

For convenience, a portion of **Figure 7C** (from the 1st to the 16th) is enlarged in **Figure 9**. Peng's DFA measures vertical differences between fitting curve and real data (**Figure 9**). Thus, Peng's DFA looks at "critical phenomena" according to physicists. But mDFA does not do those measurements: mDFA looks at 10 heartbeats at once (**Figure 9**).

Peng's DFA looks at individual heartbeat one by one (**Figure 9**). What mDFA looks at is how much [z_i] sequence has proceeded over time within a box, sometimes up and sometimes down. Therefore, mDFA can see ever changing undulation or fluctuate in each box (**Figure 9**). Fluctuation is not always stochastic noise. Rather, fluctuation carries previously unknown hidden information. It is sometimes hidden threat. It is sometimes high-risk information.

One might think that both, Peng's DFA and mDFA, have a similar calculation concept. But, mathematically, there is a gap between their concepts. According to Peng's paper and successor's publications, there is a tipping point (changing point and critical point) at the box-size11-beat. Peng et al. labeled the scaling exponent as alpha-1 and alpha-2. Alpha-1 corresponds to box-size ranges smaller than 11. In turn, alpha-2 corresponds to box-size range greater than 11. However, mDFA does not detect this tipping point. We found that box-size smaller than 30 (30-beat-box-sise) does not carry physiologically significant information. mDFA program begins computation from 10-beat-box-size and goes to 1000-beat-box-size (see **Figure 10**) but draws regression lines from box-size greater than 30 (see next section).

5.9 Scaling

Figure 10 shows typical mDFA results from heartbeat data of a crab (**Figure 10A**). mDFA makes a log-log plotting graph. Abscissa axis, which is box-size and ordinate is shown in variance (**Figure 10B**). If a clear slope can be seen in the graph, mathematically, the slope represents scaling property buried in heartbeat signal.

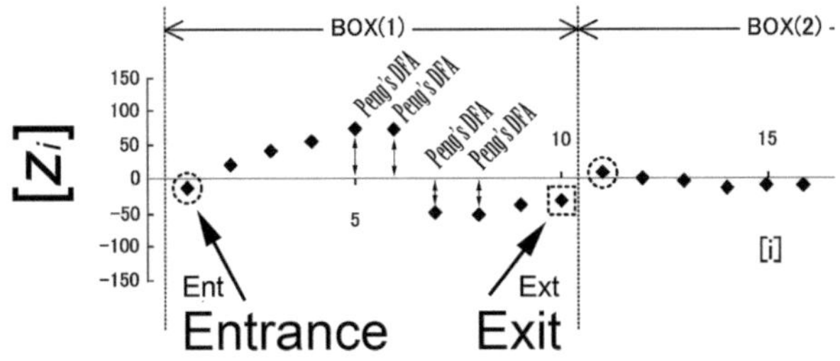

Figure 9.
Diagrammatic representation of the difference between Pang's DFA and mDFA. Pang's DFA calculates vertical difference between data and zero-line (four small arrows). Peng's DFA computes average of 10 data in a box. Total data length is 2000-beat. There are 200 boxes of 10-beat-box. Peng's DFA thus can get 2000 data in one-box-size-calculation. In turn, mDFA computes the difference between Ent and Exit (Ext – Ent). mDFA thus can get 200 data. After finishing box-10 computation, the program increases box-size and repeats calculations cyclically: box-11, box-12, box-13, and so on.

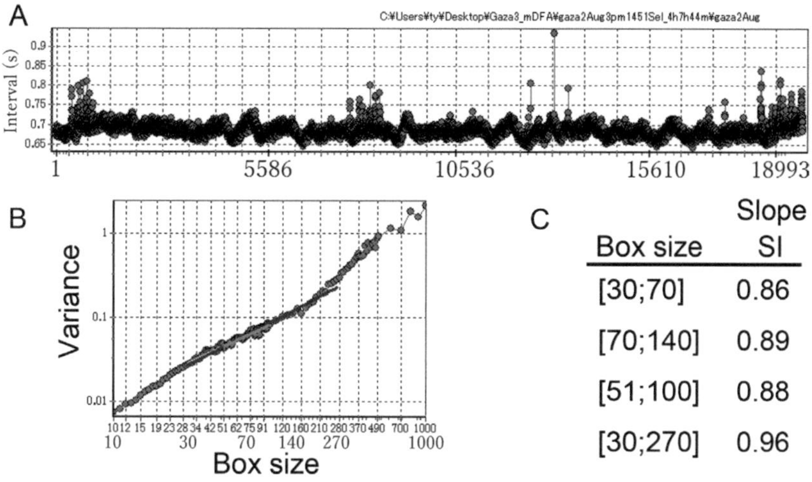

Box size	Slope SI
[30;70]	0.86
[70;140]	0.89
[51;100]	0.88
[30;270]	0.96

Figure 10.
An example mDFA. (A) Heartbeat interval time series; Gazami crab. (B) mDFA graph, box size versus variance; slope determines the SI. (C) Four box-size-ranges and corresponding SI. Crossover can be seen at two points, at a box-size of ~30 beat and ~200 beat. Physiological interpretation is under study. In this computation, 20,000-beat data were used, although 2000-beat data produce similar results (not shown).

5.10 Scaling range

By drawing a strait regression line, mDFA computes SI (**Figure 10B**). But the length of the regression line, from where to where, is unsolved. Default mDFA program draws six lines at once in a log-log graph. Each slope corresponds to the respective SI value. In **Figure 10C**, four SIs are computed. Standard six box-size ranges include: [30; 70], [70; 140], [51; 100], [30; 140], [130; 270], and [30; 270] (**Figure 11**). In our studies, unless otherwise specified, the six set is not changed for the sake of NOT to create confusion, while it can be changed infinitely. **Figure 11** is the final style after testing a variety of ranges. We have been using this "unaltered program" made by former master student, Tanaka [7] for over 10 years.

Meanwhile, those who have the skill of programing can easily make his/her own program. It is a high school level mathematics. Scaling range can be determined by the person who makes it. Other mathematical procedures, such as averaging, square root fitting, and drawing a scaling line, are not complicated tasks.

So, we can guaranty that any mDFA program surely captures cardiac scaling properties. mDFA works in physiology. Thanks to great names, William Harvey (1628 Circulation), Marcello Malpighi (1653 Medical Dr., Capillary), Ludwig Traube (1872 Alternans pulse), Willem Einthoven (1903 EKG), and Anton Julius

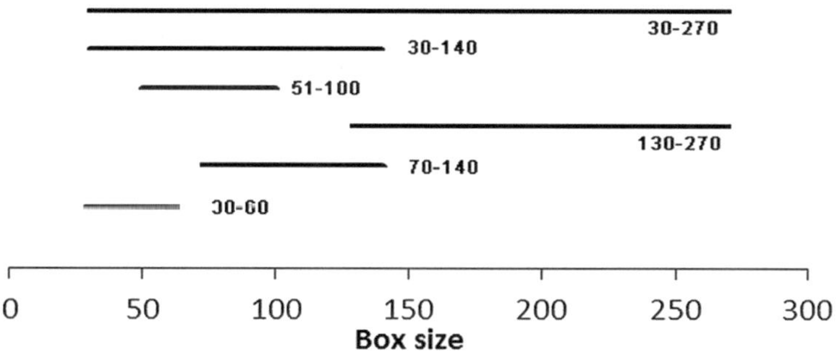

Figure 11.
Standard six box size range.

Carlson (1904 Heart physiology on model animals), for example, basics of physiology would never change forever.

If two persons have their own mDFA program, then they analyze the same data, and then one can say, "my computed SI is 1.10" and the other can say, "mine is 0.93." This kind of "contradiction" can happen. But it is NOT a big deal. We need to overlook the details. Both are around 1.

Regarding mDFA computation, please see the following sections that show what we calculate from heartbeat data.

EKG signal is generated by the cardiac system. Elements in the system are linked to each other. The system cannot work properly without feedback connections. If one can find a scaling-line in the graph (**Figure 10B**), the heart system is working properly. If a line is bending or winding, something is wrong in the body system. We guaranty so. And if a subject is healthy, mDFA tells you that the SI-value is around 1 (1.0).

As far as we know, this scaling property of the heart system was first documented in 1982 [8] and then in 1990s [9]. They proposed this nice metric theory. They used a well-known mathematical idea. We must say we moved it forward. But mDFA is based on different concepts—this is the novelty of this research—than Peng's concept as shown below. We just use the scaling property that the cardiac system inherently has.

5.11 Physiological interpretation

After finding the slope, linear fitting is necessary to determine SIs. We draw a regression line from box-size 30-beat to 270-beat as the best range for interpreting physiological meaning of heartbeat data [1]. In our study for more than 10 years, SI is "always" obtained from the regression line ranging from box-size 30-beat to box-size 270 beat (**Figure 10B**).

A 30-beat time length corresponds to about 30 s. A 270-beat time length is approximately 3–5 min. We feel sure that life prefers "3–5 min" period length: boxing round fighting time, for 3 min; hit song one musical performance, for 3 min; instant noodles cooking time, for 3 min; and a pain killer medication, coming on 3 min after taking it. We found that it seems convenient and correct that mDFA draws a line within a box-size range [30; 270] (**Figure 10C**) to check if the body system is alright or not.

6. Human general population

6.1 Ethics

We try to record EKGs of general population including people in the classroom, in the exhibition hall, company-employees, university-employees, and people at a scientific conference venue [1]. Every experimental subject was treated as per the ethical control regulations of universities (Tokyo Metropolitan University; Tokyo Women's Medical University; Universitas Advent Indonesia, Bandung; Universitas Airlangga, Surabaya, Indonesia).

6.2 SI: reproducibility

All our data are collected by the author [1, 5, 6]: invertebrate heart study since the 1980s, human data since approximately 2000, and materials data since 2010. The mDFA program was made by a former master student Tanaka [7] in about 2004.

mDFA results are reproducible and consistent. We found stratification phenomena that provide evidence for the quantitative measure SI links to various physiological phenomena in a one-to-one manner [1, 7, 10–12]. Arrhythmic heartbeat decreases SI [11]. Non-REM sleep decreases SI [12]. Premature ventricular contraction (PVC) decreases SI [11]. Alternans (harbinger of death rhythm) decreases SI [7]. Anxiety, fear, and worry decrease SI [10]. University president, vice president, president-secretary, and dean professor all have a low SI [1] but teaching-only professors have a healthy SI (SI = ~1.0) [1]. A happy content housewife has a healthy SI too [1].

Meanwhile, we encountered some healthy looking but non-healthy-heart subjects in general population [1]. We found that these subjects have had received cardiac surgery. Their myocardium is indeed injured like the crab specimen shown in **Figure 3**. All of them had a high SI. A person who has an implantable cardioverter had SI = 1.22 [1]. A person who has stent-replacement had SI = 1.26 [1]. A person who had bypass-surgery had SI = 1.38 [1]. A person who had a surgery due to ventricular septal defect had SI = 1.41 [1]. However, until today, we have never met any person, in general population, who keep maintaining a high SI and later passed away.

Ergometric exercise increases SI [1]. We think that hard exercise is probably NOT a healthy behavior for normal humans.

Heartbeat is repetitive muscle contraction. It is a cyclic behavior. It is oscillation. It is a fluctuating event. SI can quantify these unstable movements. SI can tell us the cardiac system's condition. If SI is around 1, there is no health problem regarding the heart and its control. However, if your SI is high or low, maybe I say, "Better see a doctor." But the research has only just begun.

We would like to declare that mDFA can sense warning sign although mDFA cannot identify what is wrong or what is going on.

7. Nonliving material

7.1 Introduction

We learned that mDFA detects abnormality of the heart system. Especially, we learned that a system failure increases SI up from the basic value 1.0. The failure of the heart is generally myocardial cell damage. Myocardial cells are the elementary structure of the system. Analogically, it is like a material that is made by granulated elements [13]. We expected that mDFA might contribute to nonliving system because mDFA works well in the heart.

In **Figure 3**, when a crab heart's cells were damaged by an electrode, the damage caused a significant shift of SI (toward SI = 1.5) (**Figure 3**). In human heart cases, a person who had a surgery due to ventricular septal defect, the cardiac surgery might be a major cause that pushed SI up from normal SI [1] (see a large SI, 1.41, in Section 6.2).

Materials have different properties, meaning each has its own quirks when processing [13] like the hearts.

7.2 Abnormal vibration

In nonliving material experiments, we use a piezoelectric sensor for vibration detection. It is a mechanical monitoring device made for a cardiac pulse sensor (ADInstruments, Austuraria). The sampling rate is 1 kHz in our heart experiments. The heart beats at about 1 Hz in rate. In turn, a motor rotates ~3000 times per min.

It is 50 Hz oscillation. We set the sampling rate (ADIinstruments) at 20–40 kHz in non-living material experiments. After recording vibration signal, we capture peaks and conduct mDFA as usual.

We use consecutive 2000 peaks for the analysis. We obtained vibration data lasting for about 40–60 s. The methods for both living and nonliving vibration are fundamentally the same.

7.3 Electric motor

A motor has a design that will safely operate for a long time. We realized that a running motor did not break easily [14]. We therefore covered the motor by glass wool to enhance overheat (**Figure 12**).

We monitor vibration wave travelling through the fixed base by a piezoelectric device. Vibration was analyzed by mDFA, like the heartbeat analysis. **Figure 13** shows results, which demonstrate that abnormality is captured by mDFA. We used a box-size range [30; 270] as in the heartbeat analysis (**Figure 13A**). SI is around 0 when running without overheating (see the periods designated as P–U, **Figure 13**). **Figure 14** shows an example of mDFA.

A sound "bang" occurred at the time of the end of U, and smoking started. It is hazardous. We stopped running the motor at about 10 min (**Figure 13C**). After the bang sound, one can see that SI significantly increases. The motor still ran till 10 min at the same speed although overheated. In **Figure 13C**, one can see that amplitude of signal significantly decreases. We estimate decreased stiffness and durability. It might be plastic's inherent weakness. We later opened the motor. Overheat caused softening of plastic parts inside the motor, especially plastic materials surrounding the brush. Softened plastics may absorb vibration energy more than hard-cold one. **Figure 13B** demonstrates wave patterns. U is much noisier than Q although running speed does not change. It is an induction motor (200 V, 50 Hz).

We found that a box-size range [30; 270] seems to work properly as in the heart. However, the box-size range [130; 270], which is a much narrower range, seems to contribute greatly to capture "warning sign" about "failing" motor (see # in **Figure 13A**). The plotting "130–270" indicates that the "time-window size [130; 270]" detects malfunctioning earlier than other "window sizes" (**Figure 13A**). Moreover, the SI value (see the plotting of "window-size 130–270") increases rapidly in number earlier than the "bang" sounds (**Figure 13A**). This is beyond doubt. But we must say that details are not known for explicit interpretation.

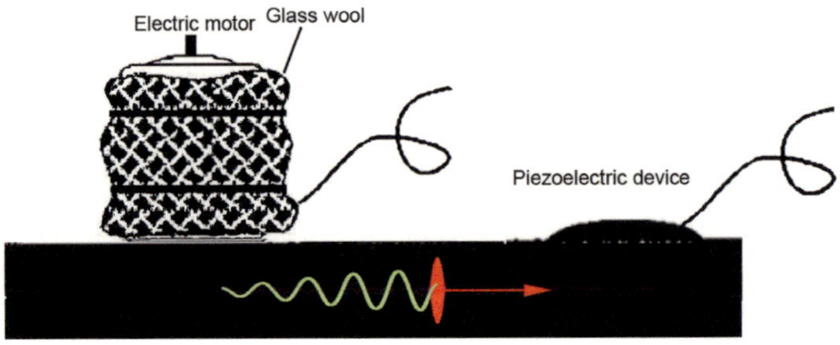

Figure 12.
Schematic image of electric motor experiment. A hand-dryer motor (in this study 100 and 200 V tested) is set on a base. A piezoelectric device monitors vibration. The sensor is connected to a logger (ADInstrument). It is the same analysis method as the EKG study, except for a higher sampling rate (20 or 40 kHz).

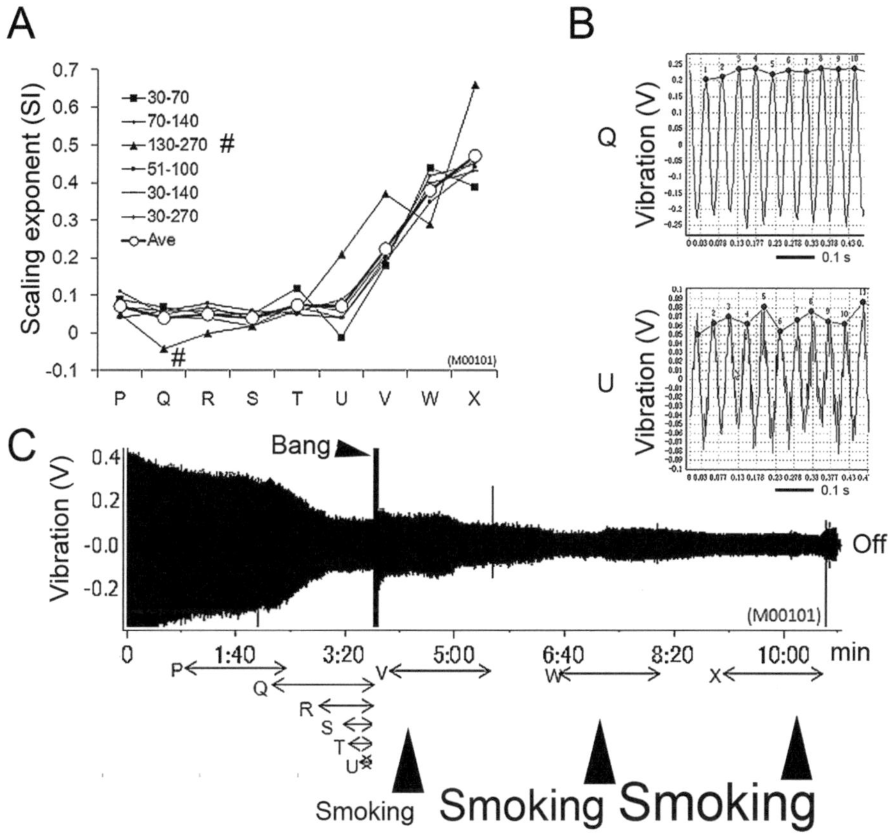

Figure 13.
A 200 V motor, overheating experiment. (A) mDFA results. (B) Example waves and peak-identifications. (C) Time zero, run motor. The time period P, heating up is insignificant. Q, Amplitude of signal decrease. Between U and V, a bang sound. Then smoking is started and gradually become worse, then the room was filled with smoke. Motor was stopped before igniting (see Off). It is perhaps fair to say that the time period lengths of R, S, T, and U are not identical. One thus can ignore R, S, T, and U, which do not interfere interpretation of data. The surge of large increase of SI occurs after Q.

7.4 Aluminum bar

We test material's toughness or the fracture toughness by changing stress intensity. **Figure 15** shows a diagrammatic image of fracture testing. We set a cantilever. A bar was tightly set on the fixed base (see W, in **Figure 15**). Vibration was applied by a speaker (see S, in **Figure 15**). Vibratory wave was monitored by a piezo device. We apply downward pressure to one end of the bar (see caption of **Figure 15**).

Figure 16 shows the summary of results. There is no load during the time period P. After P, an increase of a load is started. The bar distorts reversibly (Q, R, S in **Figure 16**). At time T, a catastrophic event, an irreversible fracture occurred (TU in **Figure 16**).

Figure 16C shows that, at normal state, SI is near 0.5 instead of near zero (**Figure 17**). This is not like motor (**Figure 13**).

Before the fracture event, **Figure 16C** shows unique results: SI attains a very high level, 1.2–1.4. This high value of SIs reminds us of ischemic heart disease's SI [1] (see Section 6.2). It is hidden threat. It is a high-risk state.

In summary, mDFA can monitor shear stress. Q-R-S periods are a period of reversible deflection. Among them, R-S periods are special. It is at a risky time: an elastic state shifts suddenly into an irreversible condition, which is catastrophe.

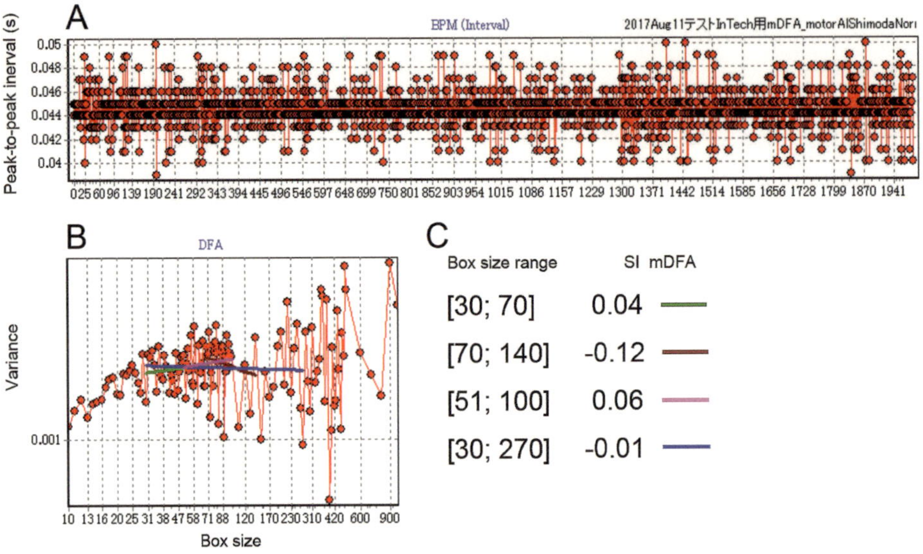

Figure 14.
An example mDFA. Running motor. (A) An interval time series. (B) mDFA graph, log-log plotting and fitting lines. (C) mDFA results. Note that slopes are vertical, meaning a normal healthy motor has an SI around 0. (Supplement note: if the testing motor is set on unstable fixed base, such as the automobile engine in the car, giving rise to a resonance with the surroundings, then SI becomes approximately 0.5, like stochastic noise. Undescribed in this article).

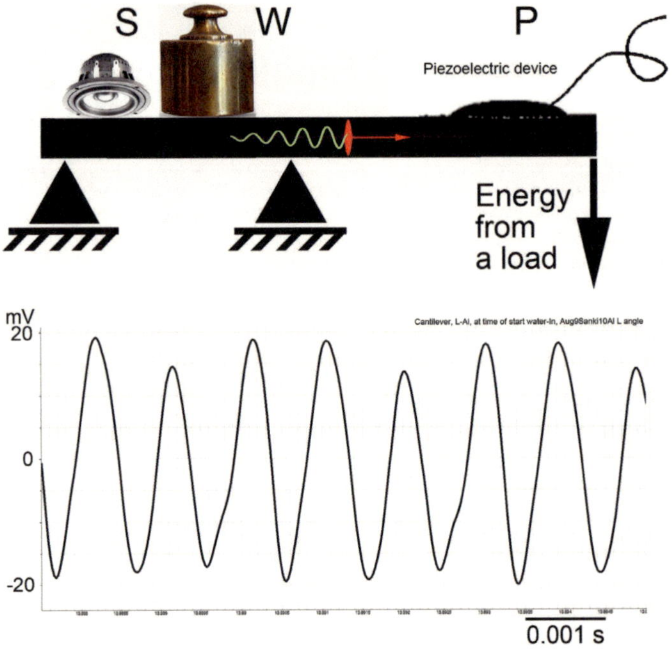

Figure 15.
Schematic image of cantilever set-up. A bar (black bar) set on a base by a weight (W). One end receives downward force accelerating at a constant speed (an arrow). A speaker (S) generates vibration. Fluctuation signal passes through the material bar and reaches to a piezoelectric sensor (P), which is connected to PowerLab 4/20 (ADInstrument, Australia). The recording method is the same as that of EKG-study as aforementioned, except for a higher sampling rate (20 or 40 kHz). Inset: PowerLab's recorded wave profile, without load.

For the safety, at the level where SI is about 0.6, inspectors are recommended to do their job for checking abnormality of materials, bridges, buildings, etc. However, it is just a biologist idea.

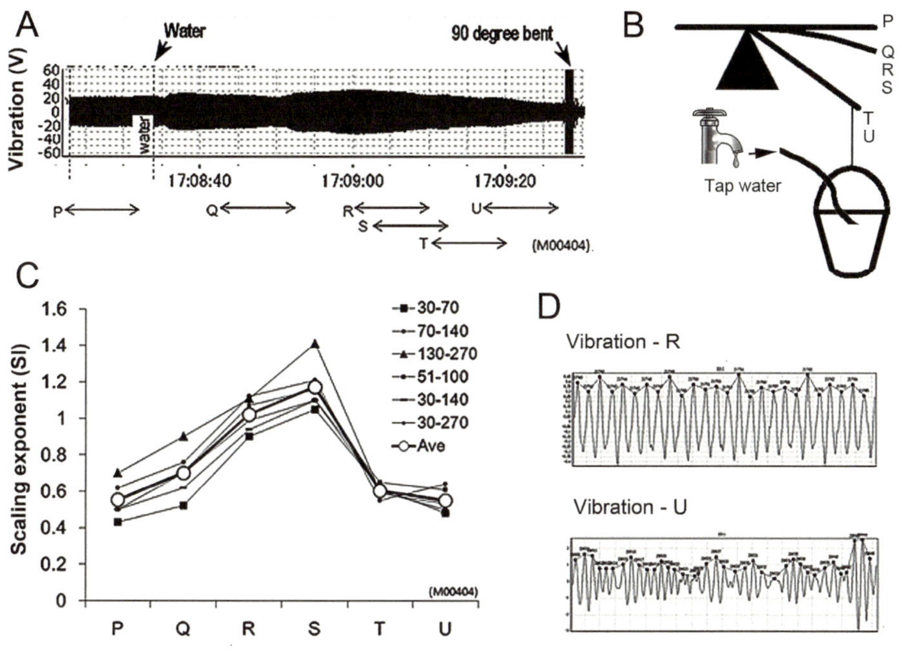

Figure 16.
mDFA results of cantilever experiment. (A) Vibration recording. (B) Diagrammatic representation of deflecting bar by a load. (C) SI value change over time. (D) Example waveforms. Material: aluminum L-shaped angle bar, cross section 3 mm thickness and 20 mm side. Load: tap water, flowing into a bucket at a rate approximately 5 L per min.

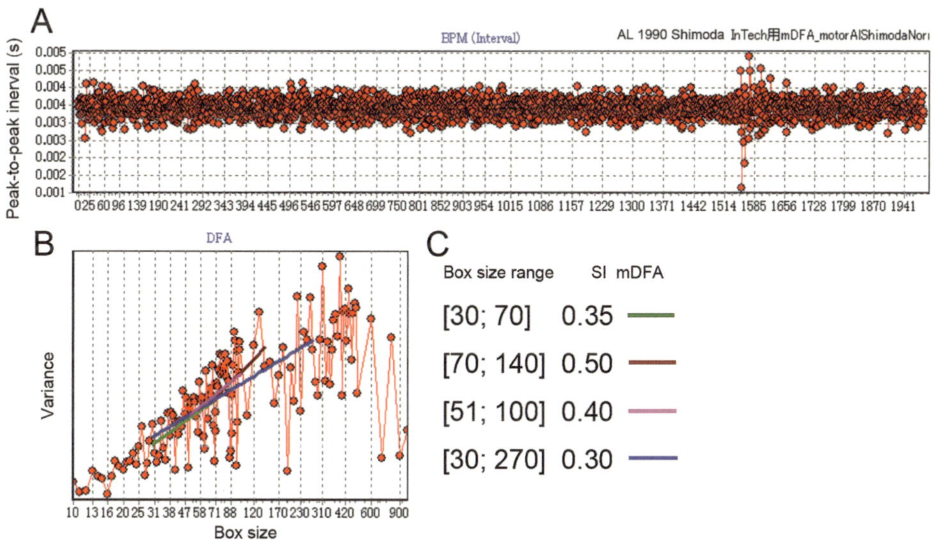

Figure 17.
An example mDFA. Aluminum bar. No load. See **Figure 14** *for comparison to motor tests.*

7.5 Earthquake

We consider that earthquake is fracture of rock structure underground, meaning fracture of materials. We expected that mDFA might help analyze these data. Ground vibration data are available from government institutions. We tested the idea.

Figure 18 shows a gigantic earthquake vibration, recorded by a seismometer at Narita in Japan, approximately 500 km away from the seismic center. The date was 11 March, 2012, afternoon.

We obtained raw seismic data (**Figure 18**) from the High Sensitivity Seismograph Network Japan. Two arrows in **Figure 18** show the time of the big event. A flat line before the big event is NOT a true straight line (see A1 in **Figure 18**). We magnified y-axis scale. The conversion discloses hidden small vibrations (A2 and A3, **Figure 18**).

Linear y-axis is inconvenient for peak detection, because some are extremely large. We converted y-axis. We plotted it in a logarithmic scale. B2 shows the square of B1 (**Figure 18B2**). Then, we make it upside down (from **Figure 18B2, B3**).

After this pretreatment, we captured peaks by a lab made program (**Figure 18C**). We use the same program for R-R peak detection in the heartbeat study. The C1-trace shows a portion of C2-trace in enlarged time scale. The C3-trace shows an example of peak-to-peak interval time series.

Figure 19 shows mDFA results. The observation period is from 3 to 29 March, 2012. The scaling exponent on 4 March was around 0.5 (SI = ~0.5) (**Figure 19B**). This means the vibration is stochastic movement. Then, SI grows up and attains a "risky" level about of 1.0. Since we are NOT specialists of seismology, we are afraid to say that it is hidden threat or high-risk state. However, in terms of chaos dynamic theory, 1.0 means the system's behavior is dynamic. It is like the heart system. It is never stable.

We know that it is too hasty to mention: this mDFA result is very similar to that of aluminum bar fracture experiment shown in **Figure 16**. In the aluminum bar fracture, SI grows up during the elasticity period, that is, reversible deflection.

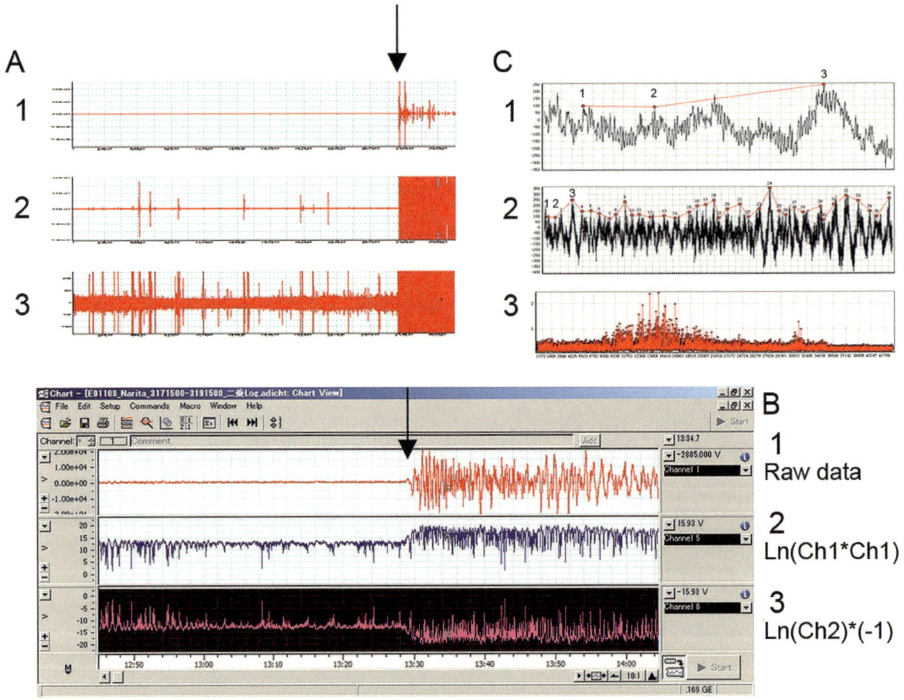

Figure 18.
Earthquake data and pretreatment for mDFA. A, Hidden vibration being exposed to view, a raw earthquake data (A1), an enlargement of Y-axis (A2), further enlargement of Y-axis (A3). B, An explanation of pretreatment procedures. A raw earthquake data (B1), logarithmic output of the square of B1, an inverse of B2 (B3). C, An example explanation of peak detection, with a faster chart speed (C1) and with a slower time (C2), accomplished peak-to-peak interval time series (C3).

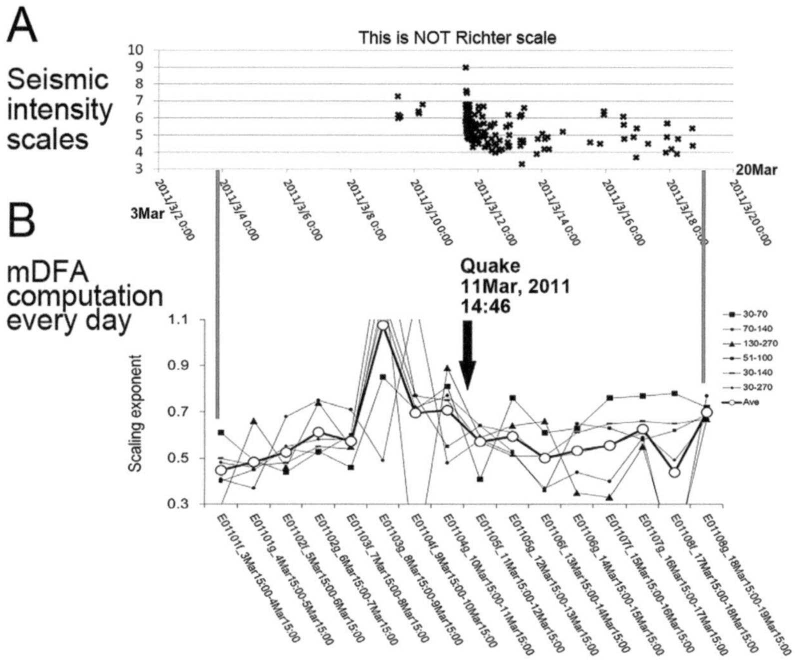

Figure 19.
mDFA results. Data: Narita seismometer. (A) Seismic intensities reported by Japan Meteorological Agency. The marks (X) indicate individual earthquakes. Five or six noticeable tremors before the big event are seen. (B) mDFA results. A plotting (marked -o-) represents averaged SI-value of all box-size ranges.

After the fracture event, aluminum's SI returns to a normal SI-value (around 0.5). **Figure 19B** shows similar trait in **Figure 16**.

This earthquake investigation using mDFA has just begun. We are not professionals, but **Figure 19B** results are remarkable. We hope that seismologists and engineers might possibly have an interest in mDFA.

8. Conclusions

mDFA computation is simple. It is high school level mathematics: first, constructing peak-to-peak interval time series [x]; second, calculating an average value x_{ave} using 2000 data; third, computing Σ $(x - x_{ave})$; fourth, cutting the time series into box; fifth, drawing a fitting biquadratic line in each box; sixth, finding the first data (Ent) and the last data (Exit) in each box; seventh, calculating the difference (Exit–Ent) in each box; eighth, calculatingΣ $(\text{Exit} - \text{Ent})^2/2000$ then obtaining "variance" (statistics of root-mean-square); ninth, changing the size of box one by one, and repeating the statistics cyclically; tenth, making a log-log plotting graph that is box-size versus variance; eleventh, drawing a linear regression line; and twelfth, measuring slope of the line. At the end, the slop denotes SI.

A 2000 "interval" data are fundamental. This length of data is not always rigid. A 2200-interval, for example, produces similar results to that of 2000-interval.

We hope that many people can make their own mDFA program. The basics are averaging, root-mean-square computation, and fitting. It has never been proposed before.

In the present study, we extended mDFA to nonlife system. We then provide the comprehensive results by analyzing various real-world data, which include oscillation/vibration generated from materials. All our results are versatile; it could be applicable to the heart, a motor, materials, and possibly earthquake motion.

The heartbeat data and/or material-vibration are not static and ever-changing phenomena. They fluctuate momentarily. We did not expect that a nonlinear-way-of-thinking method (mDFA) can distinguish the states between "intact heart" and "isolated heart" when we started investigation without questioning. It was more than what we thought. Invertebrate experiments, that is, isolated heart experiments and unpredictable death experiments were a never-to-be-forgotten experiment to discover the power of the mDFA technique.

Peng's DFA and mDFA each has different scope and concept. Peng's DFA considers criticality. In turn, mDFA deals with characteristics of fluctuation embedded in signal that fluctuates over time. In the future, not only a biologist but also engineers and seismology physicists hopefully study much more data in their discipline, by using mDFA.

If we need to find abnormality of a system, mDFA always requires comparison with a baseline SI value. There is a baseline value. That quantification method makes mDFA reliable and versatile.

Acknowledgements

This work was supported in part by the JSPS Grant 17K01364.

Author details

Toru Yazawa[1]* and Shinji Omata[2,3]

1 Tokyo Metropolitan University, Tokyo, Japan

2 Yamagata University, Yamagata, Japan

3 Symphodia Phil Co. Ltd., Yamagata, Japan

*Address all correspondence to: yazawa-tohru@tmu.ac.jp

References

[1] Yazawa T. Modified Detrended Fluctuation Analysis (mDFA). ASME Series Momentum Press. NY: LLC; 2015. Print ISBN: 9781606506127

[2] Shimizu H, Fujisawa T. Peduncle of Hydra and the heart of higher organisms share a common ancestral origin. Genesis. 2003;**36**(4):182-186. DOI: 10.1002/gene.10213

[3] Yazawa T, Kuwasawa K. The cardio-regulator nerves of the hermit crabs: Anatomical and electrophysiological identification of their distribution inside the heart. Journal of Comparative Physiology. A. 1984;**154**:871-881. DOI: 10.1007/BF00610688

[4] Yazawa T, Katsuyama T. Spontaneous and repetitive cardiac slowdown in the freely moving spiny lobster, *Panulirus japonicus*. Journal of Comparative Physiology. A. 2001;**187**:817-824. DOI: 10.1007/s00359-001-0252-z. PMID: 11800038

[5] Yazawa T, Kiyono K, Tanaka K, Katsuyama T. Neurodynamical control systems of the heart of Japanese spiny lobster, *Panulirus japonicus*. Izvestiya VUZ. Applied Nonlinear Dynamics. 2004;**1-2**:114-121

[6] Yazawa T, Katsuyama T, Katou A, Kaizaki H, Yasumatsu M, Ishiwata T, et al. Fourier spectral analysis and micro-bore column HPLC analysis of neuronal and hormonal regulation of crustacean heart. Hosei University Tama Bulletin. 2001;**16**:29-40 (In Japanese)

[7] Yazawa T, Tanaka K. Scaling exponent for the healthy and diseased heartbeat: Quantification of the heartbeat interval fluctuation. In: Sio-Long AO et al., editors. Advances in Computational Algorithms and Data Analysis. Lecture Notes in Electrical Engineering, vol. 14 Springer: Dordrecht; 2009. pp. 1-14. DOI: 10.1007/978-1-4020-8919-0.ch1

[8] Kobayashi M, Musha T. 1/f fluctuation of heartbeat period. IEEE Transactions on Biomedical Engineering. 1982;**BME-29**: 456-457. DOI: 10.1109/TBME. 1982.324972

[9] Peng CK, Havlin S, Stanley HE, Goldberger AL. Quantification of scaling exponents and crossover phenomena in nonstationary heartbeat time series. Chaos. 1995;**5**:82-87. DOI: 10.1063/1.166141

[10] Yazawa T. Chapter 2: Anxiety, worry and fear: Quantifying the mind using EKG time series analysis. In: Mohamudally N, editor. Time Series Analysis and Applications. Rijeka, Croatia: InTech Open; 2017. pp. 7-22. DOI: 10.5772/intechopen.71041

[11] Yazawa T, Shimoda Y, Katsuyama T. Chapter 42: DFA, a biomedical checking tool for the heart control system. In: Ao SL, Rieger BB, Amouzegar M, editors. Machine Learning and Systems Engineering. Dordrecht: Springer; 2010. pp. 547-556. DOI: 10.1007/978-90-481-9419-3_42

[12] Yazawa T, Shimoda Y, Hutapea AM. Evaluation of sleep by detrended fluctuation analysis of the heartbeat. AIP Conference Proceedings. 2011;**1737**(199):199-210. DOI: 10.1063/1.3627205

[13] Mogire E. What's the Difference Between Grinding Materials? Technologies. Materials. Machine Design. February 28, 2019. https://www.machinedesign.com/materials/ [Accessed: March 04, 2019]

[14] Yazawa T, Shimoda Y. Detrended fluctuation analysis: An experiment about the neural-regulation of the heart and motor vibration. In: Kim H, Ao SI, Amouzegar M, Rieger B. editors IAENG Transactions on Engineering Technologies. Lecture Notes in Electrical Engineering Book Series Vol. 247. Springer: Dordrecht; 2014. pp. 665-682. DOI: 10.1007/978-94-007-6818-5_47

Chapter 4

Countermeasure for High Level Sound Generated from Boiler Tube Bank Duct

Masaaki Mori and Kunihiko Ishihara

Abstract

Heat exchangers and boilers are widely used in various plants such as power plants and chemical plants. In the heat exchanger and the boiler, a high level sound is sometimes generated from the tube bank installed in a duct. In tube banks, warm gas flows outside of the tube, and due to the external flow around tube banks, the Karman vortex shedding occurs. At a certain frequency of the Karman vortex shedding that depends on the flow velocity, the resonance phenomenon called the self-sustained tone occurs. The self-sustained tone might cause noise problems in the surroundings, losses due to plant shutdown, etc. For suppression of the self-sustained tone, insertion of baffle plates in the tube bank is generally used. Although the insertion of baffle plates has been adopted for a long time, it is difficult to decide the effective insertion positions. On the other hand, a method using perforated plates has been proposed for the suppression of the self-sustained tones. In this chapter, we review the generation mechanism of the self-sustained tone clarified experimentally and numerically, and the methods for suppressing a self-sustained tone using baffle plates and perforated plates.

Keywords: self-sustained tone, high level sound, acoustic resonance, lock-in phenomenon, countermeasure, baffle plate, perforated plate

1. Introduction

Heat exchangers and boilers are widely used in various plants such as power plants and chemical plants. In heat exchangers such as boilers and gas heaters, a high level sound is sometimes generated and it results in a serious problem such as a plant shutdown or non-operation. A high level sound is generated in tube banks installed in a duct. In tube banks, water flows inside of tubes, warm gas flows outside of tubes, and Karman vortex shedding occurs. The vortex shedding frequency depends on the flow velocity. In contrast, a resonance frequency called an acoustic natural frequency, which is independent of the flow velocity, is determined by the duct size and the sound speed. When the two frequencies coincide, a resonance phenomenon occurs at a certain velocity [1–9]. Ziada and Oengören [10] have shown that vortex excitation results from the formation of periodic vortices in the space between tubes by visualization experiments in the waterway. Hamakawa et al. [11] focused on effect of arrangement of tube banks, and investigated the characteristics of vortex shedding and acoustic resonance from in-line and staggered tube banks.

When a resonance phenomenon occurs at a certain velocity, if the acoustic damping is small, a high level sound continues as flow velocity increases. This phenomenon is called the self-sustained tone [12, 13]. The self-sustained tone might cause the surrounding noise problem, and also cause plant shutdown and, hence, production losses, etc.

For a countermeasure of the self-sustained tone, a method of inserting a partition plate called a baffle plate inside the duct is generally used. In this method, the baffle plate inserted inside the duct is assumed to increase the natural frequency of the duct, detune the frequency of the vortex shedding from the tube bank and the acoustic natural frequency of the duct, and suppress the resonance phenomenon [14–16]. However, Ishihara et al. [12] demonstrated that the natural frequency of the duct decreases by inserting the baffle plate, and a decision of an appropriate insertion position of the baffle plate is not easy. Hamakawa et al. [16] have investigated the effect of the baffle plate on the acoustic resonance generation from in-line tube banks with small cavity, and they clarified that although sound pressure level of an acoustic mode perpendicular to the flow (lift mode) is suppressed by a baffle plate, that of an acoustic mode parallel to the flow (drag mode) increases. Ishihara and Takahashi proposed that flexible walls such as rubber boards are set on the duct walls for suppressing the self-sustained tone [17]. They expected that the vibration of the flexible walls damp the lift resonance mode when self-sustained tone is generated. They demonstrated that the suppression effect of the rubber sheet appeared when the tension of it is small and it is located at just the tube bank and downstream of the tube bank. On the other hand, to suppress the self-sustained tones, a method using perforated plates and cavities has been proposed by Ishihara and Nakaoka [13]. A perforated plate has long been used in various noise-control applications, such as vehicle exhaust systems, ducts, hearing protection devices, and acoustic panels, because it is well known that perforated plates have an acoustic damping effect [18–20]. Ishihara and Nakaoka [13] thought that a resonance mode perpendicular to the flow (lift mode) might be suppressed by a damping effect of perforated plates, when the self-sustained tone occurred.

In this chapter, we review the generation mechanism of the self-sustained tone clarified experimentally and numerically, and the methods for suppressing a self-sustained tone using baffle plates and perforated plates.

2. Generation mechanism of self-sustained tone

2.1 Self-sustained tone

The Karman vortex shedding frequency *fv* is generally proportional to the flow velocity and the natural frequency of the duct *fa* is constant value determined by the duct size and the sound speed. When the flow velocity increases, the *fv* approaches *fa*. Before the *fv* reaches the *fa*, the vortex shedding frequency suddenly locks on to the natural frequency of the duct. The resultant high level sound occurs at or nearly at the natural frequency of the duct, and this phenomenon is called a lock-in or lock-on. There are many studies on the excitation mechanism of a lock-in or lock-on phenomenon in the tube bank [1–11, 21, 22].

The relation between frequency and flow velocity in a lock-in phenomenon is represented in **Figure 1**. A high level sound called a self-sustained tone occurs due to a lock-in phenomenon [12, 13]. In a lock-in phenomenon, as shown in **Figure 1**, the frequency slightly rises as the flow velocity increases. Furthermore, the lock-in occurs at a certain flow velocity, and does not occur in accordance with the large acoustic damping of the duct if the flow velocity increases. However, with the small

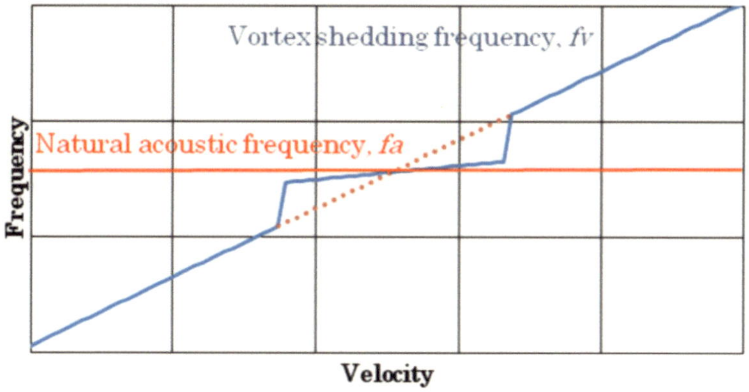

Figure 1.
Relation between frequency and velocity in the case of lock-in phenomenon.

acoustic damping of the duct, the lock-in continues as the flow velocity increases, and the sound pressure level remains high [13]. **Figure 2** shows that when the shedding frequency of the strong vortices generated in the tube bank almost coincides with the acoustic resonance frequency of the duct, the strong sound field in the duct is excited. As a result, the vortices and the sound field in the duct cause the strong interaction. This phenomenon is a self-excited mechanism.

Also, focused on the self-excited acoustic resonance of two side-by-side cylinders in a duct, the mechanism of the self-excited acoustic resonance is investigated by experiments and numerical solutions [21, 22]. It was found that dynamic lift fluctuation on the cylinders and strong in-phase vortex shedding synchronization are generated by the acoustic resonance. Shahab Khushnood et al. reviewed and summarized the flow-induced vibrations and acoustic resonance in heat exchanger tube bundles [23].

2.2 Setup of experiment

Ishihara et al. [12, 13] performed the experiments to investigate the self-sustained tone. **Figure 3(a)** and **(b)** represents the setup of the experiment and the tube bank. The duct is made of acrylic plates that have a thickness of 1 cm. The tube bank consists of an array of bronze tubes whose diameter is D = 6 mm. The array geometry is represented in **Figure 3(b)**, where the spacings T/D and L/D are

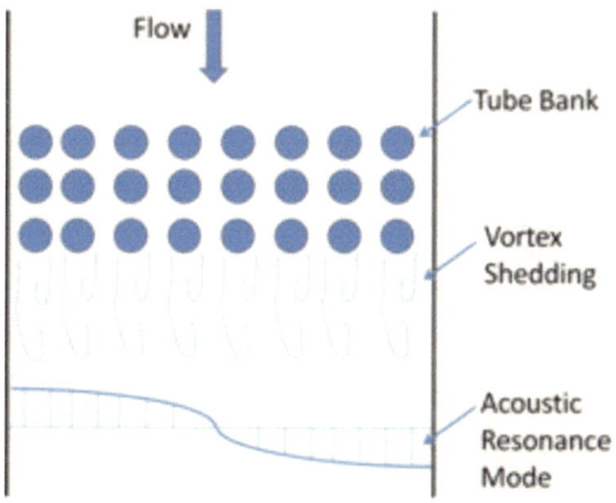

Figure 2.
Acoustic resonance and lock-in phenomenon.

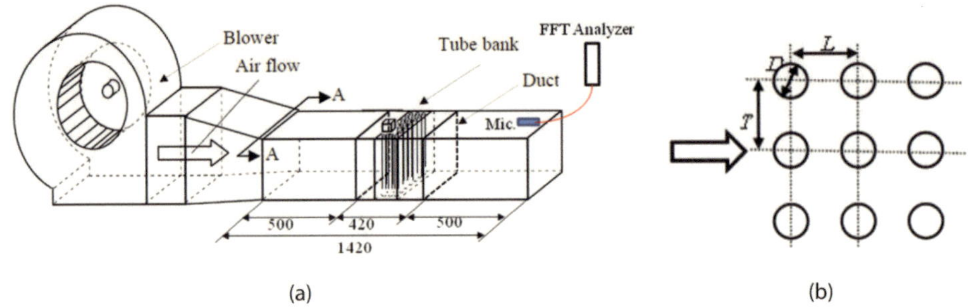

Figure 3.
Setup of experiment and array geometry of tube bank [24]. (a) Setup of experiment. (b) Array geometry of tube bank.

2.0. In the tube bank, there are 9 rows of tubes in the flow direction and 19 tubes in the width direction, which is perpendicular to the flow, and the length in the flow direction is 102 mm. The sound pressure signal is measured using the microphone set near the duct outlet as shown in **Figure 3(a)**, and converted to frequency domain with FFT analyzer. The flow velocity is changed by controlling the rotational speed of the blower using the inverter. A high level sound is generated when the shedding frequency of vortices generated in the tube bank nearly coincides with the acoustic natural frequency of the duct system. A flow velocity measurement hole was provided at a position 125 mm upstream of the tube group, and the flow velocity (U) was also measured by using a hot wire anemometer. The gap flow velocity V_g is obtained from the continuous equation, defined by the flow velocity U and the ratio of the area of the duct outlet to the area of the tube bank clearance which is 234 mm (Duct width)/(234 mm – 19 (number of tubes in the width direction)*6 mm (tube diameter)) = 1.95, and represented by $V_g = 1.95\ U$. The measurement frequency ranges from 100 to 2000 Hz. The sound pressures signal is measured with the sampling frequency of 10,000 Hz, the number of averages of 1000, and the frequency resolution of 20 Hz.

2.3 Results of experiments

The sound pressure spectrum at each gap velocity (11.4, 15.7, 19.6, and 21.3 m/s) is shown in **Figure 4**. As shown in **Figure 4**, the self-sustained tone is slightly generated at V_g = 19.6 m/s, and is clearly generated at V_g = 21.3 m/s. The peak frequency of the self-sustained tone is 740 Hz. The relation between overall sound pressure level and the gap velocity obtained from the experiments is represented in **Figure 5**. Sound pressure level generated by a flow in a duct generally follows the 5–8th power laws [13], and the sound pressure level generated in the duct in this experiment follows the 5th power law. Sound following the 5th power law is the ordinary aerodynamic sound. The sound pressure level rises as the gap velocity increases by following the 5th power law when the self-sustained tone is not generated (when the gap velocity is lower than 20 m/s). However, the sound pressure level is over 100 dB when the self-sustained tone is generated (when the gap velocity is higher than 20 m/s). **Figure 5** represents that when the gap flow velocity is over 22 m/s, the overall sound pressure level remains high and is over 110 dB [15]. The self-excited tone is generated at the gap velocity over 20 m/s, and it indicated that if the Strouhal number St is assumed to be 0.22, the vortex shedding frequency f_v is $f_v = St * V_g/D = 0.22 * 20/0.006 = 733$ Hz. Meanwhile, the resonance frequency f_a in the width direction of the duct is $f_a = C/2L = 340/2/0.234 = 726.5$ Hz.

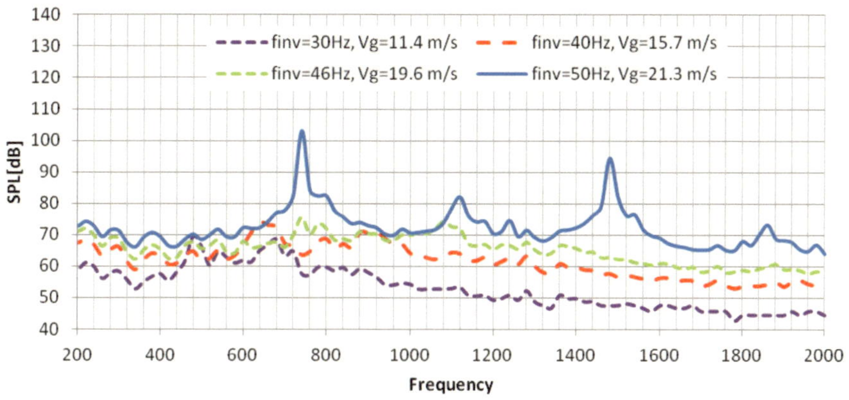

Figure 4.
Spectra of sound pressure level.

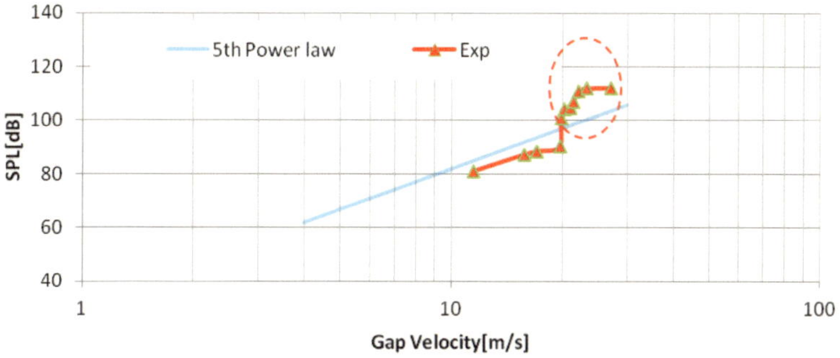

Figure 5.
Relation between overall sound pressure level (200–2000 Hz) and the gap velocity.

Taking the combined mode in the longitudinal direction into consideration, the resonance frequency f_a is obtained from Eq. (1) and equals 736.3 Hz, which is very close to the frequency of excitation at 740 Hz obtained from the experiments.

$$f_a = \frac{c}{2}\sqrt{\left(\frac{1}{l_x}\right)^2 + \left(\frac{m}{l_y}\right)^2} = \frac{340}{2}\sqrt{\left(\frac{1}{0.234}\right)^2 + \left(\frac{1}{1.42}\right)^2} = 736.3\ \text{Hz} \tag{1}$$

Here, l_x and l_y denote the longitudinal duct length and the duct width, respectively.

2.4 Unsteady CFD simulations

Mori et al. [24] performed compressible CFD simulations to capture the self-sustained tone and compare the simulation results with the measurements [13]. They confirmed that the self-sustained tone at the acoustic mode in the width direction of the duct occurs, and the sound pressure level does not follow the 5th power law when the gap velocity is high, as in the experiments. Unsteady flow fields in the duct are simulated in the paper. Inflow velocities are U = 5.846, 7.026, 8.051, 8.564, 9.590, 10.923, and 13.846 m/s, and correspond to the gap velocities, V_g = 11.4, 13.7, 15.7, 16.7, 18.7, 21.3, and 27.0 m/s, respectively. **Figure 6** represents the CFD model that is a three-dimensional computational domain. Reynolds number Re_D is based on the gap velocity V_g, and ranges from 4600 to 10,800.

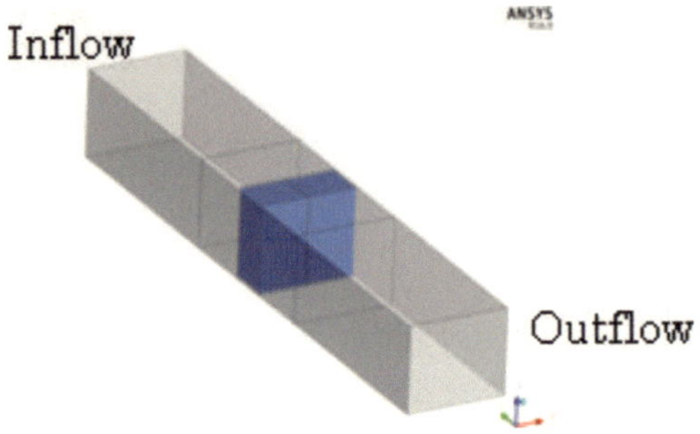

Figure 6.
CFD model [24].

Unsteady flow fields are calculated using the commercial CFD code ANSYS Fluent version 17.0. An implicit pressure-based coupled solver with second-order numerical accuracy in both space and time and compressible LES (Dynamic Smagorinsky model) calculation features have been applied. The interaction between the flow and acoustic fields need to be solved when the resonance or self-sustained tone is generated, a high level sound is generated, and the monitor point is near the noise source region. Therefore, the acoustic pressure is directly extracted from the unsteady compressible CFD simulations [25].

The origin of the Cartesian coordinate is placed at the center of the inflow boundary. The cell spacing adjacent to the wall is 0.00025 m. In the wake region near the tube bank, the cell spacing is about 0.002 m. In the far wake region, the cell spacing is stretched to 0.006 m. The domain contains 4,944,100 cells and 5,156,304 nodes. CFD simulation conditions are shown in **Table 1.**

Steady-state simulations were performed using Spalart-Allmaras (S-A) turbulence model and then used as initial conditions of transient LES simulations. The time step size corresponds to the non-dimensional time step based on f_v, 0.00733. To convert the acoustic pressure time histories obtained from CFD simulations into the frequency spectra, the discrete Fourier transform (DFT) has been applied. The acoustic pressure is extracted from 2500 steps (from t = 0.05 to 0.1 s). The sampling period is 2e-5 s.

Instantaneous snapshots of vorticity fields at $Z = 0$ plane are shown in **Figure 7(a)** and **(b)**, for the cases of $V_g = 11.4$ m/s and $V_g = 21.3$ m/s. A vortex street is formed downstream by vortices shed in the tube bank. At $V_g = 21.3$ m/s, the strength of the vortices is larger than at $V_g = 11.4$ m/s. Instantaneous snapshots of static pressure fields are represented in **Figure 8. Figure 8** shows that the value of the static pressure on the upstream side of the tube bank is larger than that on the downstream side, and distinguishing the sound pressure from the static pressure seems to be difficult. Thus, the fluctuation pressure is defined as follows to distinguish easily between the sound pressure and the static pressure [24].

$$dp = p_s - p_{mean} \tag{2}$$

Inflow boundary conditions	Steady velocity [m/s]: 11.4, 13.7, 15.7, 16.7, 18.7, 21.3 and 27.0 Temperature [K]:300
Outflow boundary conditions	Zero pressure and Non-reflective boundary condition
Wall conditions	No-slip and Adiabatic
Time step size [sec]	1e-5
Calculation Steps	10000

Table 1.
CFD simulation conditions.

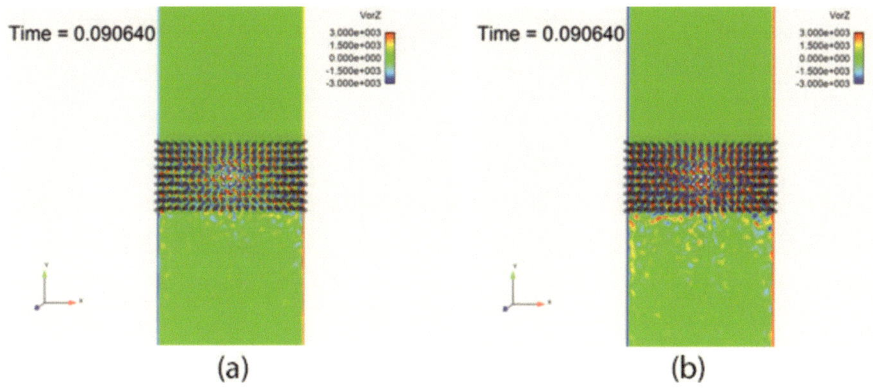

Figure 7.
Vorticity fields. (a) V_g = 11.4 m/s. (b) V_g = 21.3 m/s.

Here, p_s denotes the static pressure which is defined by $p_s = p - p_0$, p_{mean} denotes the time-averaged pressure, and p_0 denotes the ambient pressure. Instantaneous snapshots of fluctuation pressure field at Z = 0 plane obtained from the unsteady CFD simulations are represented in **Figure 9**. The value of the fluctuation pressure at V_g = 21.3 m/s is much larger than at V_g = 11.4 m/s. At V_g = 21.3 m/s, the pressure fluctuation clearly represents the resonance mode in the duct width direction.

The frequency spectra of SPL are monitored on the wall of the duct near the outflow boundary, and represented in **Figure 10**. The self-sustained tone is generated when the gap velocity V_g is 21.3 m/s, and the self-sustained tone is not generated when the gap velocity V_g is 11.4 m/s. The peak frequency of the self-sustained tone is about 740 Hz and its high harmonic frequency, 1480 Hz, when the gap velocity V_g is 21.3 m/s. This frequency, 740 Hz, is close to the resonance frequency in the duct width direction, 726.5 Hz, and that in the combined mode in the longitudinal direction, 736.3 Hz, as mentioned in Section 2.3. The resonance frequencies obtained from the theory (without the flow) and the CFD simulations or the experiments are slightly different because of the absence or presence of the flow. **Figure 1** shows that the slight increase of the resonance frequency occurs with an increase of the gap flow velocity [24]. The predicted SPL of the dominant tone at 740 Hz reasonably agrees with the measured one. The generation of the higher harmonic at 1480 Hz, which is surrounded by the red circle in **Figure 10**, is also predicted as in the experiments. The predicted SPL of the dominant tone, which is assumed to contribute most to the overall SPL when the self-sustained tone is generated, reasonably agrees with the measured one.

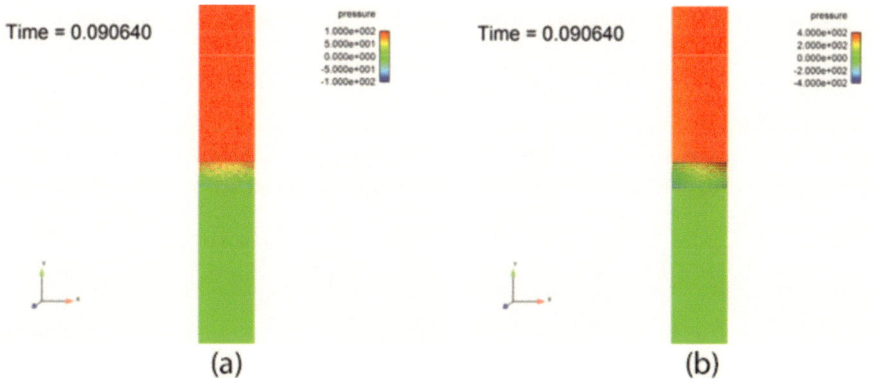

Figure 8.
Static pressure fields. (a) V_g = 11.4 m/s. (b) V_g = 21.3 m/s.

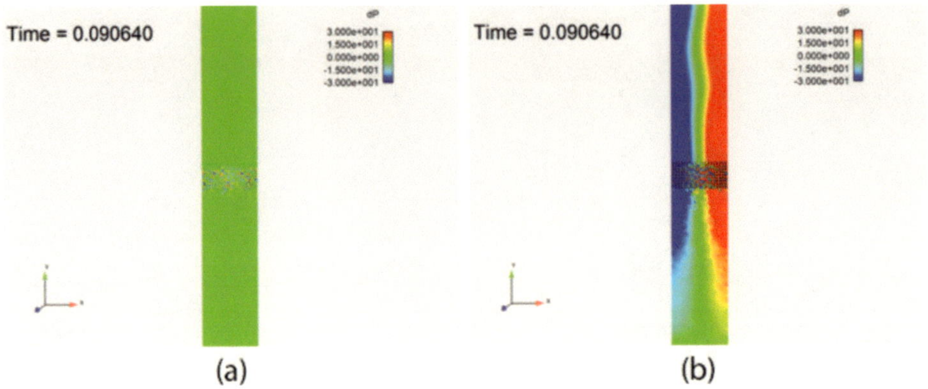

Figure 9.
Fluctuation pressure fields. (a) V_g = 11.4 m/s. (b) V_g = 21.3 m/s.

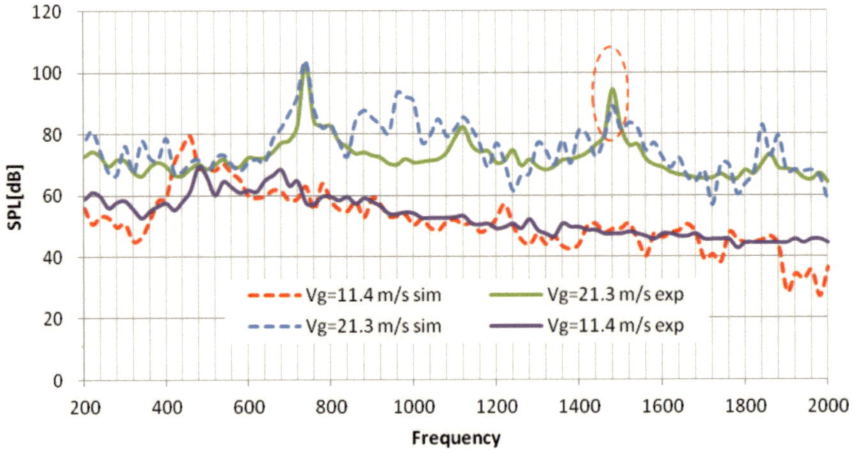

Figure 10.
Spectra of sound pressure level.

The relation between overall sound pressure level and the gap velocity obtained by both the simulations and experiments is represented in **Figure 11**. In both the simulations and experiments, when the gap velocity is low and below 20 m/s, the sound pressure level rises as the gap velocity increases by following the 5th power law. However,

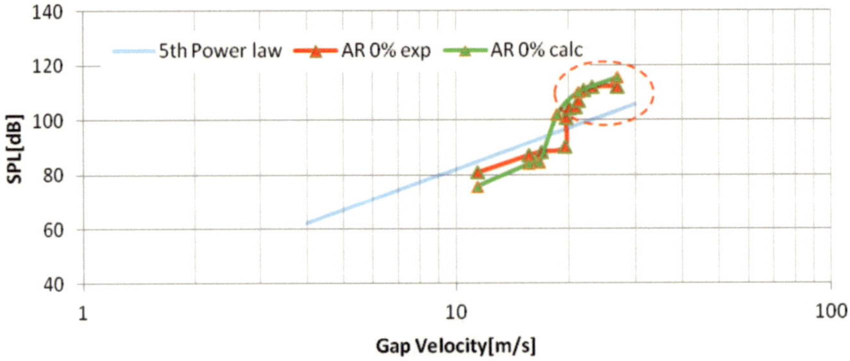

Figure 11.
Relation between overall sound pressure level (200–2000 Hz) and gap velocity.

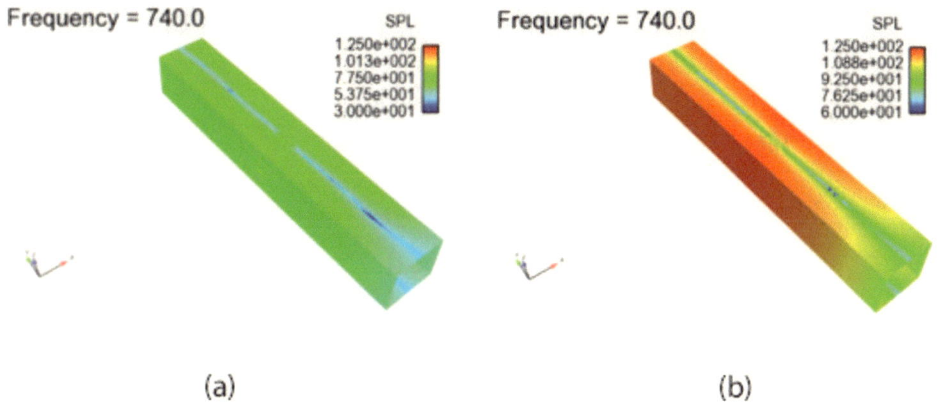

Figure 12.
SPL on the wall of the duct ([24]). (a) V_g = 11.4 m/s. (b) V_g = 21.3 m/s.

when the gap velocity is higher than 20 m/s, the self-sustained tone is generated and the sound pressure level is high and over 100 dB not by following the 5th power law. **Figures 10** and **11** represent that the sound pressure levels obtained by the simulations reasonably agree with those obtained by the experiments. **Figure 12** shows the SPLs on the wall of the duct at 740 Hz; these are extracted from the unsteady CFD simulations using DFT. As in the experiments, **Figure 12(b)** shows that when V_g = 21.3 m/s (the gap velocity is higher than 20 m/s), the SPL on the duct wall clearly represents the acoustic resonance mode in the duct width direction or the combined mode in the duct width and longitudinal directions. However, when V_g = 11.4 m/s (the gap velocity is lower than 20 m/s), the acoustic resonance mode is not clearly represented as shown in **Figure 12(a)**. The fluctuation pressure field shown in **Figure 9(b)** is consistent with the acoustic mode at 740 Hz shown in **Figure 12(b)**, which means that the dominant mode in the self-sustained tone is the acoustic mode in the duct width direction or the combined mode in the duct width and longitudinal directions, and is close to the acoustic mode obtained from the Eq. (1) and the experiments.

As shown in **Figures 10** and **11**, the simulations show a reasonable agreement with the experiments in terms of the generation prediction of the self-sustained tone.

3. Countermeasure for self-sustained tone using baffle plate

3.1 Setup of experiment

In this section, we describe the experiments performed by Ishihara et al. [12, 26]. They have investigated the appropriate insertion position of the baffle plate for the suppression of the self-sustained tone and the mechanism of suppressing the self-sustained tone by inserting the baffle plate. The setup of the experiment is shown in **Figure 3(a)** and the duct used in this experiment is shown in **Figure 13**. The tube bank consists of an array of bronze tubes whose diameter is D = 6 mm. The array geometry is represented in **Figure 3(b)**, where the spacings T/D and L/D are 2.0. In the tube bank, there are 5 rows of tubes in the flow direction and 18 tubes in the width direction, which is perpendicular to the flow, and the length in the flow direction is 60 mm. The sound pressure signal is measured using the microphone set near the duct outlet as shown in **Figure 3(a)**, and converted to frequency domain with FFT analyzer. One tube bank is installed in the duct in the present experiment as shown in **Figure 13**. The baffle plate with a length of 60 mm, thickness of 3 mm, and height of 200 mm is inserted in the center of the tube bank

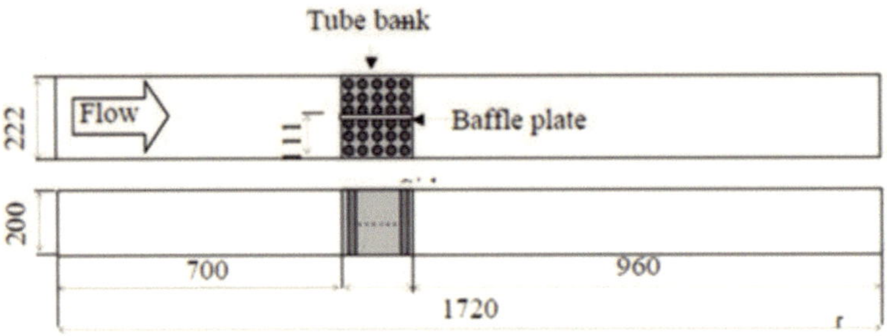

Figure 13.
Setup of experiment ([12]).

as shown in **Figure 13**. The position of this baffle plate is varied in the flow direction in increments of a row pitch (12 mm) in the tube array. The patterns of the baffle plate position are shown in **Figure 14**. An effective position of the baffle plate for suppressing the self-sustained tone is derived from this experiment.

3.2 Results of experiments

3.2.1 Natural frequency and peak frequency of self-sustained tone

Figure 15 represents the natural frequency of the duct and the peak frequency of the self-sustained tone. The vertical axis shows the frequency while the horizontal axis shows the pattern of the baffle plate positions as shown in this figure. The natural frequency of the duct can be obtained by the speaker test that was performed using the setup of the experiment shown in **Figure 13**. The peak frequency of the self-sustained tone was obtained by the ventilation experiment. The symbol Δ shows the peak frequency in the case "with baffle plate". In this case, the self-sustained tone was not generated when the baffle plate positions are −1, 0, and + 1. Therefore, we cannot see the symbol Δ in these positions. The natural frequency of the duct corresponds to the peak frequency of the self-sustained tone as represented in **Figure 15**. For a countermeasure of a self-sustained tone, a method involving the insertion of a baffle plate in the duct is generally adopted to suppress the self-sustained tone. This method is based on the idea that the baffle plate can prevent the resonance within the range of the usage flow velocity by introducing a new partition, thus increasing the natural frequency

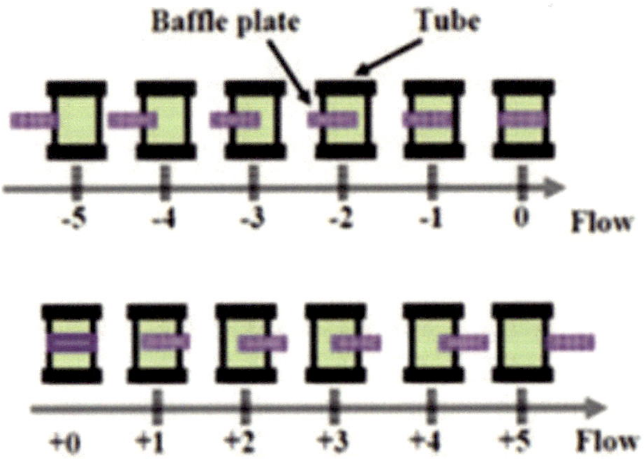

Figure 14.
Pattern of baffle plate positions ([12]).

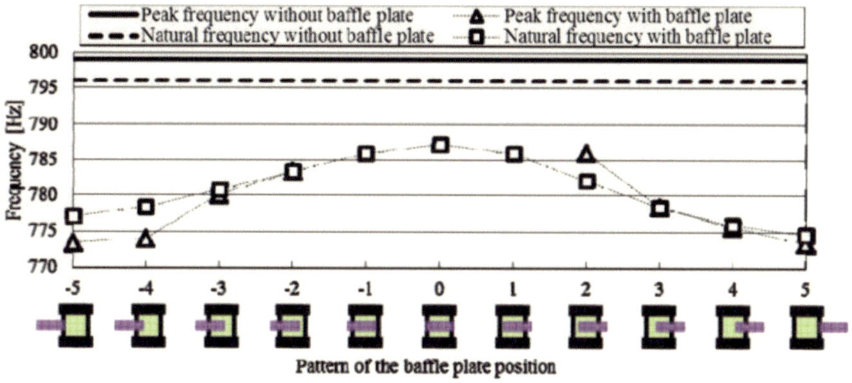

Figure 15.
Natural frequency and peak frequency ([12]).

of the duct [14, 15]. However, **Figure 15** represents that the natural frequency of the duct decreases by the insertion of the baffle plate regardless of the position in **Figure 15**, because the baffle plate cannot divide into two parts of the acoustic field of the duct due to a small length. If the baffle plate length is the same with the length of the duct, then the natural frequency becomes higher and doubles.

3.2.2 Onset gap velocity of self-sustained tone

Figure 16 represents the onset gap velocity of the self-sustained tone. The vertical axis shows the gap velocity of the tube bank when the self-sustained tone is generated and the horizontal axis shows the pattern of the baffle plate positions as shown in this figure. In patterns (−1, 0, and + 1) where the baffle plate is inserted in the entire tube bank, the self-sustained tone was not generated within the range of the flow velocity that the setup of experiment can produce. The self-sustained tone is not generated because vortices are assumed to become very small in patterns (−1, 0, +1) as described later. Furthermore, the onset gap velocity of the self-sustained tone shows a significantly different tendency between the upstream and the downstream positions of the baffle plate.

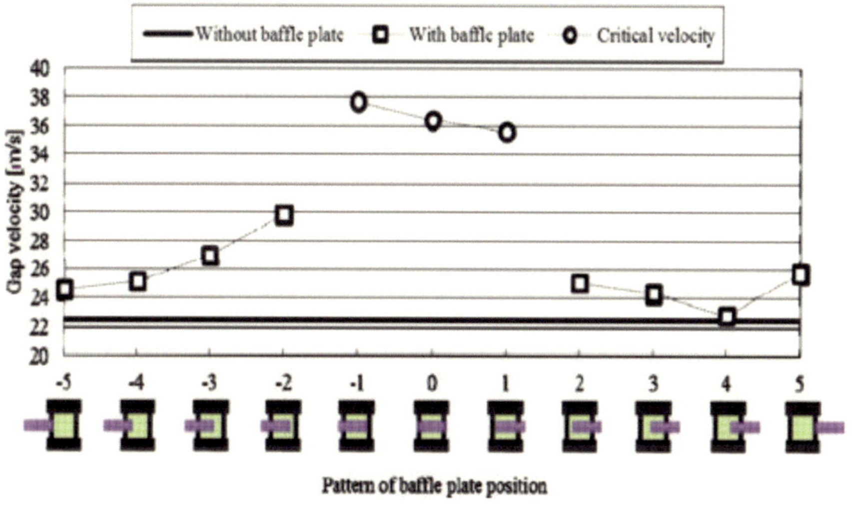

Figure 16.
Gap velocity and pattern of baffle plate positions ([12]).

The measurement position of the fluctuation velocity in the tube bank by the hot wire anemometer is shown in **Figure 17**. Because the baffle plate is inserted at the center of the tube bank, the hot wire probe is inserted in the neighboring flow channel. Additionally, the fluctuation velocity is measured between each tube row (12 mm interval). Measurement examples (the measurement position is 36 mm) of the fluctuation velocity U_f of the flow in the tube bank and SPL are shown in **Figures 18** and **19**. The vertical axis shows the fluctuation velocity of the flow and the horizontal axis shows the frequency. Two peaks (one sharp and the other dull) are represented in **Figures 18** and **19**. The sharp peak is due to the self-sustained tone and the dull peak is due to Karman vortex shedding as represented in **Figure 18**. The Karman vortex occurs in the tube bank, and the Strouhal number is about 0.13–0.19. The peak frequency of Karman vortex shedding is in proportion to the flow velocity and increases as the flow velocity increases. On the other hand, the peak frequency of the self-sustained tone coincides with the natural frequency of the duct, the sharp peak is assumed to correspond to the flow fluctuations related to the generation of the self-sustained tone. To distinguish the sharp peak from other peaks, it is referred to as "Excitation flow fluctuation" in this chapter. However, Karman vortices are generated and the excitation flow fluctuation is not generated if the baffle plate inserted in the entire tube bank suppresses the self-sustained tone.

3.2.3 Suppression mechanism of self-sustained tone by baffle plate

The sound power which the vortices add to the acoustic field of the duct is given by Eq. (3) from Howe [27]. The parameters are W: sound power [W], ρ: gas density [kg/m^3], $\vec{\omega}$: vorticity [rad/s], $\vec{U}$: flow velocity, [m/s], and $\vec{\varphi}$: particle velocity [m/s].

$$w = \rho \int (\vec{\omega} \times \vec{U}) \cdot \vec{\varphi} dV \tag{3}$$

The particle velocity in the duct is given by the gradient of the sound pressure. Moreover, the phase of the particle velocity to the sound pressure progresses by 90 degrees. The particle velocity is therefore the maximum at the node of the acoustic pressure. In addition, the particle velocity is the largest in the center of the duct width. Karman vortices are strong in the tube bank, and that means the vorticity is large in the tube bank. In addition, the vorticity strongly depends on the fluctuation velocity U_f of the flow in the tube bank. Each parameter is controlled by inserting the baffle plate in the center of the duct where these two parameters (the particle velocity and vorticity) are large. As shown **Figure 12(b)**, the sound pressure level is the maximum on the duct wall and the sound pressure level is the minimum in

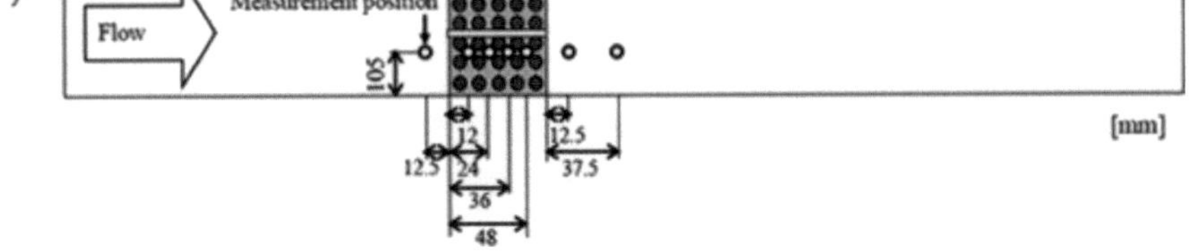

Figure 17.
Measurement positions of fluctuation velocity in duct ([12]).

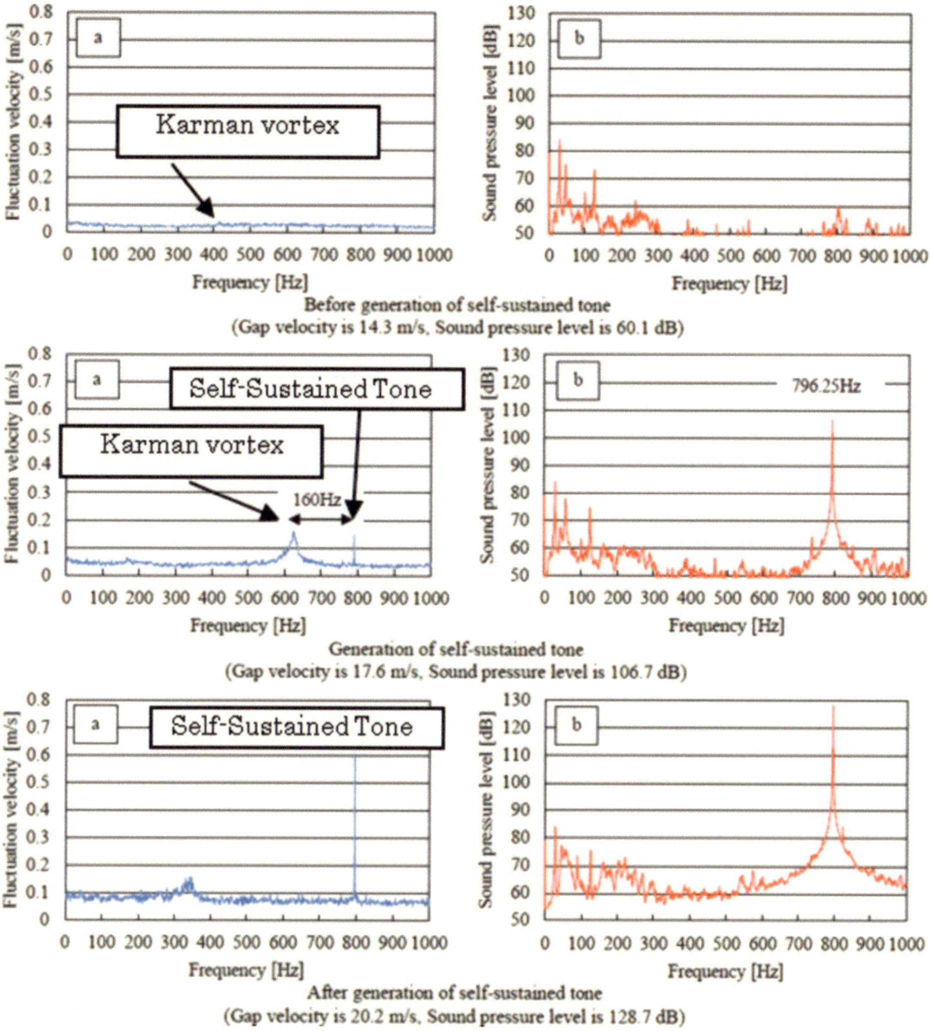

Figure 18.
The fluctuation velocity of flow and the sound pressure level at observation point ③ without the baffle plate ([26]).

the center of the duct, which means that the gradient of the sound pressure or the particle velocity is the maximum in the center of the duct and tube bank. Therefore, the vorticity and particle velocity decrease due to the baffle plate inserted in the duct; as a result, the sound power decreases. This is the suppression mechanism of the self-sustained tone.

The distribution of the excitation flow fluctuation in the tube bank when the self-sustained tone is generated is examined. Here, it has been non-dimensionalized as shown in Eq. (4) because the excitation flow fluctuation is a value depending on the flow velocity.

$$u = U_f / V_g \tag{4}$$

Figure 20 represents the distribution of the excitation flow fluctuation in the tube bank. The vertical axis shows the baffle plate positions. The circle shows the dimensionless excitation flow fluctuation and its radius indicates the value of the excitation flow fluctuation while the horizontal axis shows the measurement

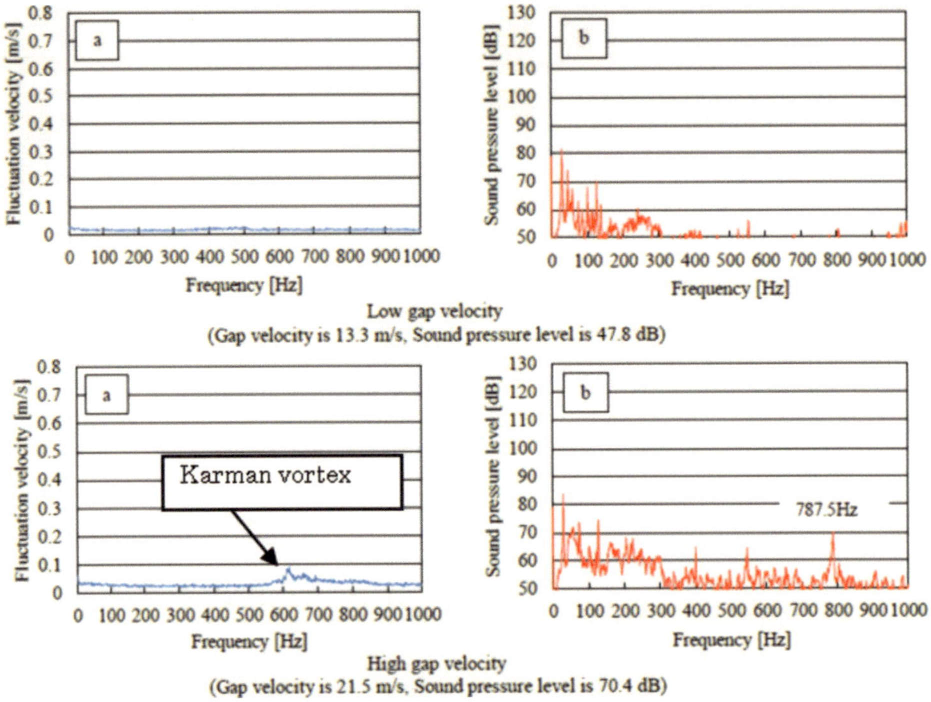

Figure 19.
The fluctuation velocity of flow and the sound pressure level at observation point ③ with the baffle plate (Pattern of baffle plate position is "0") ([26]).

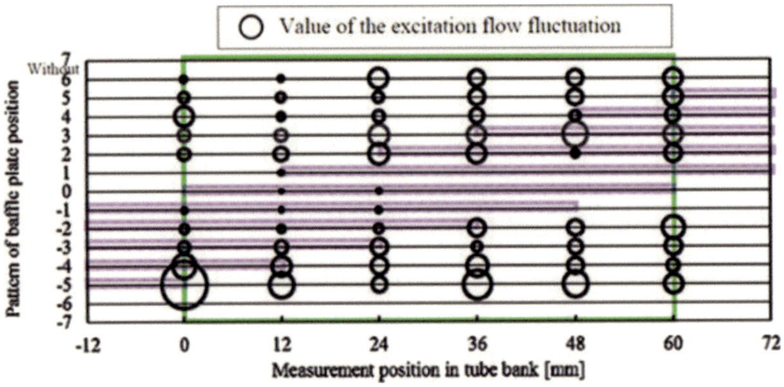

Figure 20.
Fluctuation velocity of flow on tube bank and the measurement position ([12]).

position of the flow fluctuation velocity in the tube bank. As represented in **Figure 20**, the excitation flow fluctuation is not generated in the entire tube bank under the condition where the self-sustained tone is not generated. On the other hand, **Figure 20** represents that it is generated in the entire tube bank under the condition of the self-sustained tone being generated. Therefore, the two parameters particle velocity and the excitation flow fluctuation are controlled by inserting the baffle plate. Ishihara et al. [12] thought that it is the suppression mechanism of the self-sustained tone to decrease the sound power by controlling these two parameters.

4. Countermeasure for self-sustained tone using perforated plate

4.1 Setup of experiment

In this section, we describe the experiments performed by Ishihara and Nakaoka [13] and Ishihara [28]. They carried out some experiments to examine the suppression effect of the perforated plates and cavities installed, and confirmed the suppression effect. They defined the aperture ratio ϕ of the perforated plate and investigated how the change in the value of the aperture ratio ϕ of the perforated plate affects the self-sustained tone in the duct. They varied the value of the aperture ratio ϕ from 1 to 32%. The setup of the experiment is shown in **Figure 21**. The duct is made of acrylic plates that have a thickness of 1 cm. The tube bank consists of an array of bronze tubes whose diameter is D = 6 mm. The array geometry is represented in **Figure 3(b)**, where the spacings T/D and L/D are 2.0. In the tube bank, there are 9 rows of tubes in the flow direction and 19 tubes in the width direction, which is perpendicular to the flow, and the length in the flow direction is 102 mm. The sound pressure signal is measured using the microphone set near the duct outlet as shown in **Figure 3(a)**, and converted to frequency domain with FFT analyzer.

The perforated plate is made of iron, and has a length of 400 mm, a height of 250 mm, and a thickness of 2.3 mm. A hole with a diameter of 3 mm was opened in a staggered arrangement on a plate. As shown in **Figure 22(a)**, perforated plates can be mounted from the slit (shown in green), and the duct has two cavities with a depth of L_c = 100, 66 and 33 mm. In this experiment, in order to examine the influence of the aperture ratio of the perforated plate on the self-sustained tone suppressing effect, as shown in **Figure 22(b)**, assuming a hole diameter of 3 mm, Ishihara and Nakaoka [13] made six patterns (1, 2, 4, 8, 16, and 32%) of the perforated plates. Here, the aperture ratio ϕ is the ratio of the area of the holes to the total area of the perforated plate and is defined by Eq. (5).

$$\phi = \frac{n_h\left(\frac{\pi d^2}{4}\right)}{S_p} \tag{5}$$

Here, n_h is the number of the hole, Sp = 200 × 420 mm is the total area of the perforated plate, and d is the hole diameter (3 mm). Even at the same aperture ratio, if the hole diameter is different, the influence on self-sustained tone may be different. However, in this study, it is assumed that the hole diameter is constant

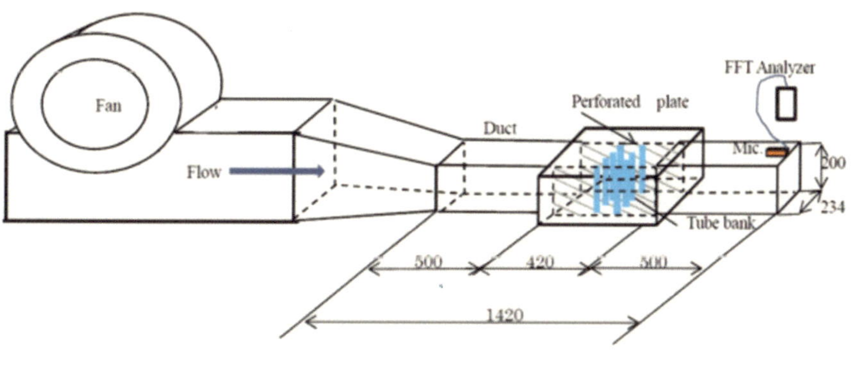

Figure 21.
Setup of experiment [29].

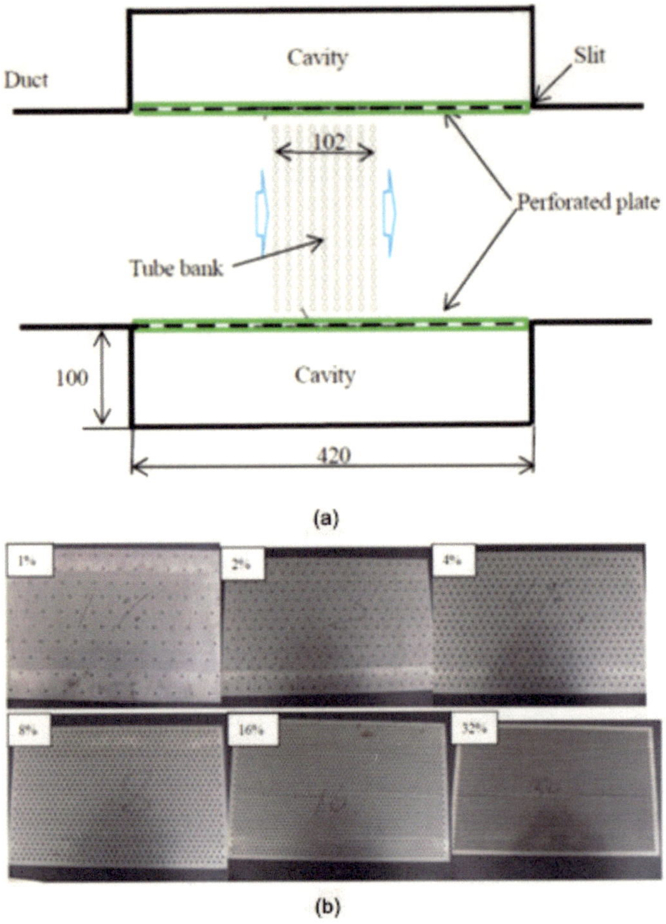

Figure 22.
Tube bank part with perforated plates ([29]). (a) Detail of tube bank part with perforated plates and cavities. (b) Perforated plate.

(3 mm) [13]. Since the case that the aperture ratio is 0% (holeless plate) is defined as the standard of this experiment, ϕ = 0% is also a parameter of the aperture ratio.

4.2 Results of experiments

An effect of the aperture ratio on sound pressure level spectra at the gap velocity V_g = 21.3 m/s is represented in **Figure 23**. The sound pressure decreases more than 30 dB in cases where the perforated plates and cavities are installed ($\phi \geq 0\%$), and the sound pressure level decreases in the high-frequency region as the aperture ratio increases. Hence, the perforated plate is assumed to have a sound-absorbing property at high frequencies. The experimental results of the relation between overall sound pressure level and the gap velocity in cases of aperture ratios ϕ = 1, 2, 4, 8, 16, and 32% are represented in **Figure 24**. Sound following the 5th power law is the ordinary aerodynamic sound, as mentioned in Section 2.3. On the other hand, as shown by the blue circle in **Figure 24**, the noise which does not follow the power law and becomes extremely large is referred to as the self-sustained tone. As shown in **Figure 24**, the self-sustained tone is generated only in the duct with the plate of the aperture ratio ϕ = 0%, which is the normal duct without holes. When the perforated plate with aperture ratios of ϕ = 1–32% is applied on the duct wall surfaces, the overall sound pressure level rises as per the 5th power law, and the self-sustained tone is not generated.

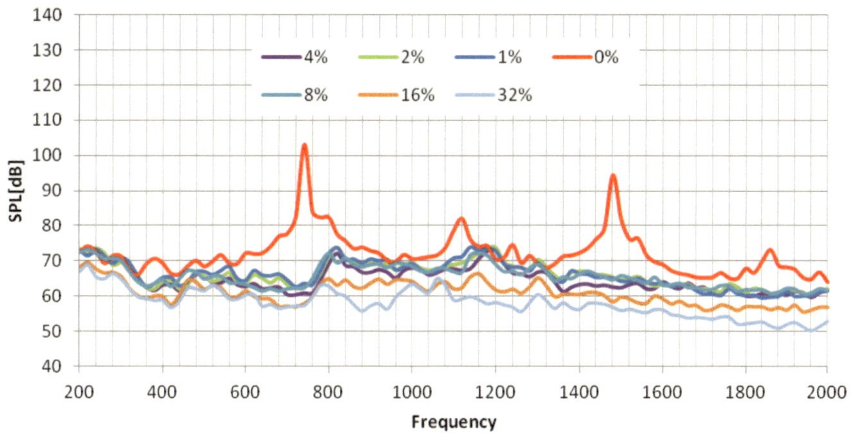

Figure23.
Effect of aperture ratio on sound pressure level spectra at V_g = 21.3 m/s.

Ishihara [28] studied experimentally the effect of a cavity volume which is used with perforated plates on the SPL. He concluded that the effect of a cavity volume on the SPL is a little. **Figure 25** shows the effect of a cavity volume or depth on the sound pressure level in the case of the aperture ratio ϕ = 4% as the typical example of the aperture ratio. Further, Ishihara [28] clarified experimentally the critical and optimum aperture ratios. The results show that the critical aperture ratio is about 0.25%, and the optimum aperture ratio is concluded to be 4%.

4.3 Unsteady CFD simulations

Mori et al. [29] performed compressible CFD simulations and acoustic simulations, and compared the simulation results with the measurements [13] to numerically verify the effect of the aperture ratio of the perforated plate on the self-sustained tone and acoustic resonance frequencies. The numerical method for unsteady CFD simulations is described in Section 2.4. The CFD model for the normal duct without holes, which corresponds to the duct with the perforated plates of the aperture ratio ϕ = 0%, is shown in **Figure 6**. As an example of the CFD model for the duct with the perforated plates, **Figure 26(a)** represents the CFD model

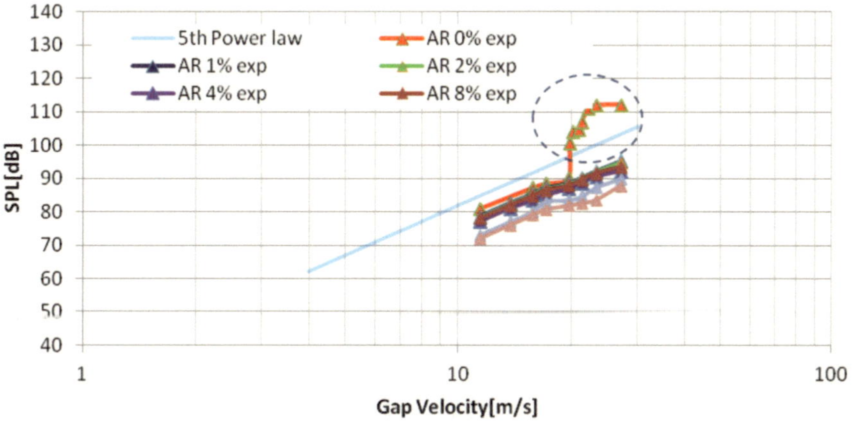

Figure 24.
Relation between overall sound pressure level and the gap velocity in cases of aperture ratios ϕ(AR) = 0, 1, 2, 4, 8, 16, and 32%, and L_c = 100 mm.

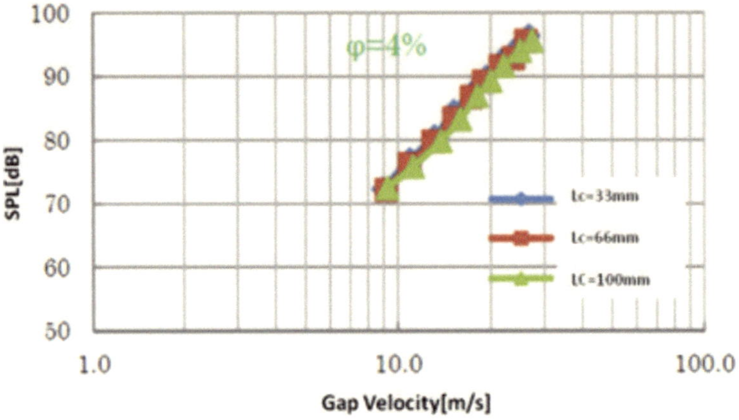

Figure 25.
Relation between overall sound pressure level and the gap velocity in cases of aperture ratios ϕ = 4% and L_c = 100, 66, and 33 mm [29].

for the duct with the perforated plates of the aperture ratio ϕ = 2%. **Figure 26(b)** shows a part of the CFD model near the perforated plates and cavities in the case of the aperture ratio ϕ = 2%. The CFD domain contains 10,727,088 cells and 6,612,429 nodes for the case of the aperture ratio ϕ = 1%; 12,053,432 cells and 7,308,716 nodes for the case of the aperture ratio ϕ = 2%; and 13,646,420 cells and 8,004,142 nodes for the case of the aperture ratio ϕ = 4%, respectively.

Figure 27 shows instantaneous snapshots of the fluctuation pressure field. Comparing the cases of ϕ = 0% with ϕ = 1%, the fluctuation pressure in the case of ϕ = 0% is much larger than that in the case of ϕ = 1%. In the cases of ϕ = 1%, the pressure fluctuation does not represent the resonance mode in the duct width direction which appears in the case of ϕ = 0% in **Figure 27(a)**.

The effect of the aperture ratio on the frequency spectra of SPL monitored on the wall of the duct near the outflow boundary is represented in **Figure 28(a)**. For comparison, both the simulated and measured data are represented, and the frequency spectra of SPL in the case of ϕ = 0% is also displayed in **Figure 28(a)**. In both simulations and experiments, the self-sustained tone is generated in the case of ϕ = 0%; however, the self-sustained tone is not generated in the cases of ϕ = 1%. The effect of the aperture ratio on the overall sound pressure level is represented in

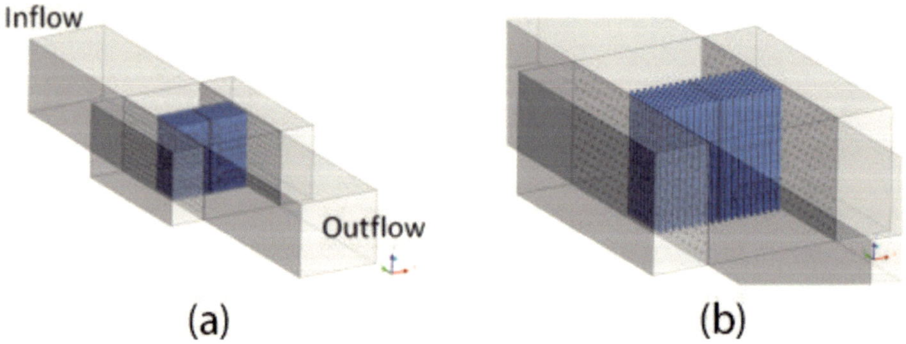

Figure 26.
CFD model for duct with perforated plates and cavities with a depth of L_c=100 mm. (a) Duct with the perforated plates of the aperture ratio ϕ = 2%. (b) Duct near the perforated plates and cavities in the case of the aperture ratio ϕ = 2% [29].

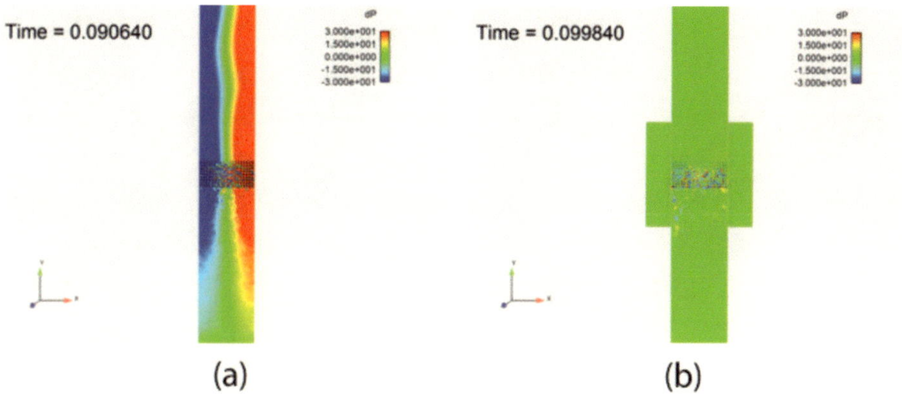

Figure 27.
Fluctuation pressure fields in the cases of aperture ratio at V_g = 21.3 m/s. (a) ϕ =0%. (b) ϕ = 1%.

Figure 28(b). The SPL decreases with an increase of the aperture ratio, and rapidly decreases when the aperture ratio is between 0 and 1%.

Figure 29 shows the SPL on the wall of the duct in frequency domain at the peak frequency, 740 Hz. In the case of ϕ = 0%, the SPL on the wall of the duct clearly represents the acoustic mode in the duct width direction as represented in **Figure 27(a)**. On the other hand, in the cases of ϕ = 1%, **Figure 29(b)** shows that

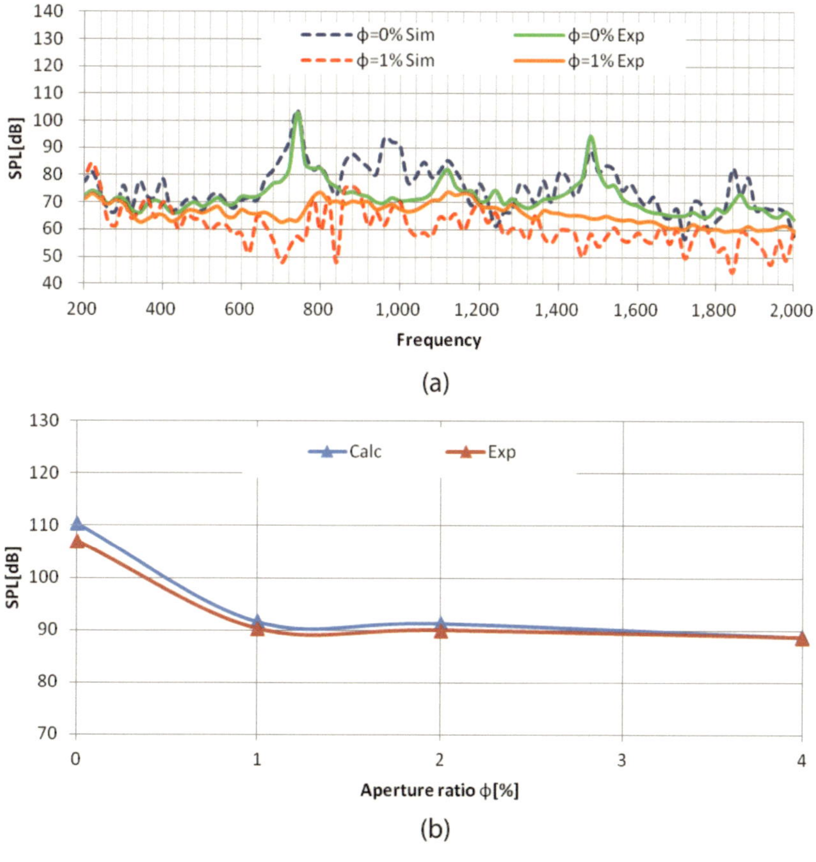

Figure 28.
Effect of aperture ratio on SPL at V_g = 21.3 m/s. (a) ϕ = 1%. (b) Overall SPL (200–2000 Hz).

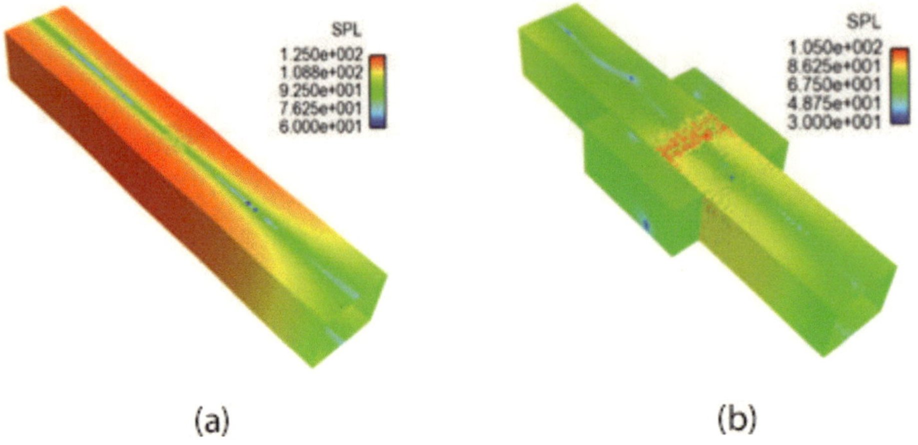

Figure 29.
SPL on the wall of the duct at Vg = 21.3 m/s [29]. (a) ϕ = 0%. (b) ϕ = 1%.

the acoustic mode is not clearly represented and the value of SPL on the duct wall is close to 40 dB smaller than that in the case of ϕ = 0%.

The relation between overall sound pressure level and the gap velocity in the cases of ϕ = 0 and 1% is represented in **Figure 30**. As represented in **Figure 30**, the self-sustained tone is generated in the case of ϕ = 0%, and the sound pressure level does not follow the 5th power law when the gap velocity is high, as mentioned in Section 2.4. However, in the case of ϕ = 1%, the self-sustained tone is not generated, and regardless of the gap velocity, the overall sound pressure level rises along the 5th power law in both simulations and experiments. Therefore, when the perforated plate is installed on the duct wall surfaces, the self-sustained tone is not generated in the CFD simulations, as in the experiments [13].

4.4 Acoustic simulations and suppression mechanism of self-sustained tone by perforated plates

The acoustic characteristics of the duct with the perforated plates and cavities without the flow were calculated by means of BEM (the commercial code, WAON) [30]. Boundary element models for the cases of the aperture ratio ϕ = 0 and ϕ = 2% are shown in **Figure 31(a)** and **(b)**, respectively, and in the case of ϕ = 2%, the duct

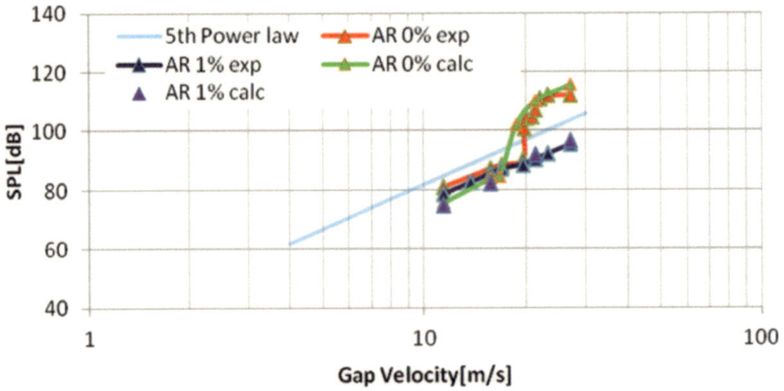

Figure 30.
Relation between overall sound pressure level (200–2000 Hz) and gap velocity in cases of aperture ratios ϕ = 0% and ϕ = 1%.

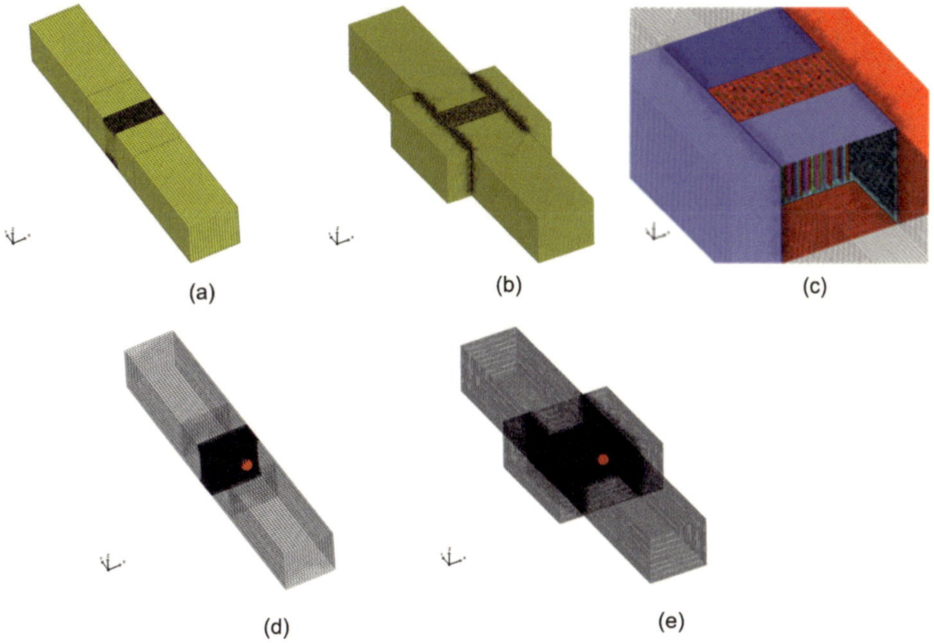

Figure 31.
Boundary element model and position of monopole point source [29]. (a) ϕ = 0%. (b) ϕ = 2%. (c) ϕ = 2%. (d) ϕ = 0%. (e) ϕ = 2%.

with the perforated plates and cavities is modeled as shown in **Figure 31(c)**. There are 80,530 boundary elements for the case of the aperture ratio ϕ = 0%; 167,361 boundary elements for the case of the aperture ratio ϕ = 1%; 226,079 boundary elements for the case of the aperture ratio ϕ = 2%; and 307,747 boundary elements for the case of the aperture ratio ϕ = 4%, respectively. The maximum element size is 0.013 m. An impedance boundary condition is imposed at the inflow boundary to consider acoustic waves moving from the inflow boundary to the outside and the value of the impedance is ρc. The outflow boundary is the surface connecting the duct inside and outside. At the outflow boundary, the interface boundary, where the particle velocity and acoustic pressure of the internal and external sound fields of the duct are coupled, is imposed. At other wall boundaries except the holes, weak absorption boundaries, whose absorption coefficient is 0.02, are imposed. The absorption coefficient "0.02"

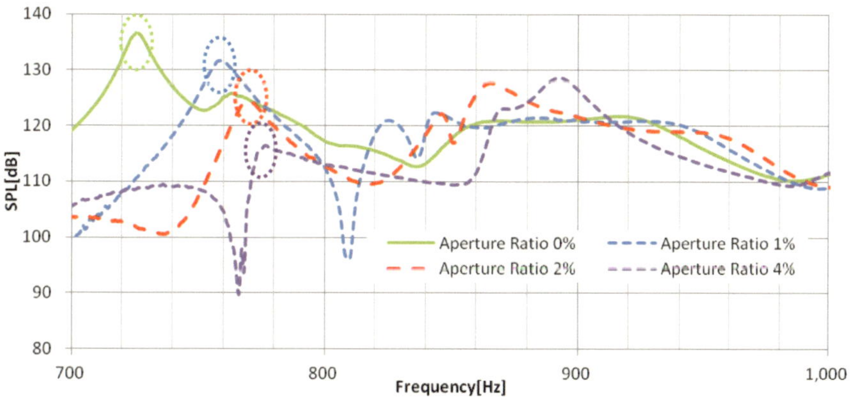

Figure 32.
Acoustic frequency responses.

Method, Aperture Ratio	*Resonance Frequency*
Theory (Aperture Ratio 0 %, without flow)	726.5 Hz
BEM (Aperture Ratio 0 % without flow)	725 Hz
BEM (Aperture Ratio 1 % without flow)	758 Hz
BEM (Aperture Ratio 2 % without flow)	769 Hz
BEM (Aperture Ratio 4 % without flow)	774 Hz

Table 2.
Peak frequency in each case of the aperture ratio [29].

corresponds to that of the general acrylic surface. To clarify the acoustic characteristics of the tube bank duct, the acoustic frequency responses have been calculated using the monopole point sources (without the flow) that are supposed to be vortices generated on the downstream side behind the tube as shown in **Figure 31(d)** and **(e)**. The magnitude of the point sources is 1 Pa at all frequencies. The point source is located at (0.25*H*, −3.86*H*, 0) downstream of the tube bank, and *H* is 200 mm in the height of the tube bank duct. In order to excite the resonance mode in the duct width direction, the sound source was arranged asymmetrically with respect to the YZ plane.

Figure 32 represents the acoustic frequency responses that have been calculated using the monopole point source (without the flow). The monitor point is located

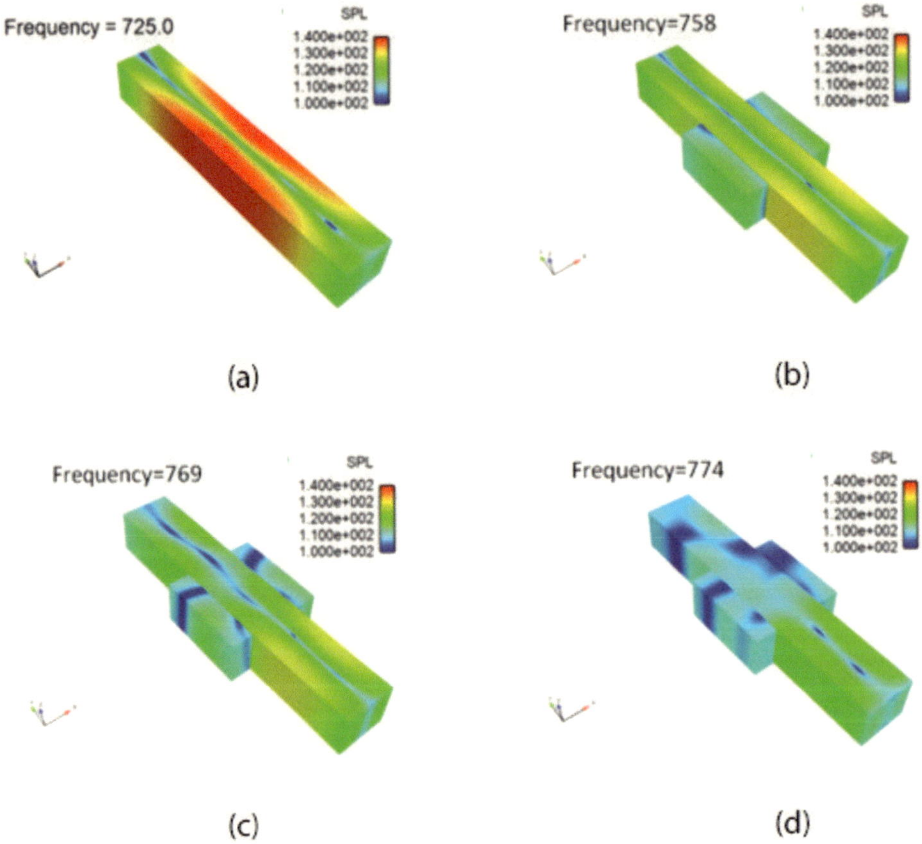

Figure 33.
Acoustic modes at each peak frequency [29]. (a) ϕ = 0%. (b) ϕ = 1%. (c) ϕ = 2%. (d) ϕ = 4%.

at $(0.585H, -4.76H, 0)$ downstream of the tube bank. In the case of $\phi = 0\%$, the peak frequency surrounded by the green circle is 725 Hz, and close to the resonance frequency in the duct width direction, as mentioned in Section 2.3. On the other hand, in the cases of $\phi = 1$, 2, and 4%, peak frequencies surrounded by blue, red, and purple circles are close to the resonance frequency in the duct width direction, and change to higher frequencies than that in the case of $\phi = 0\%$ with an increase of the aperture ratio. **Table 2** represents the peak frequency in each case of the aperture ratio. It has been also confirmed that the resonance frequency increases as the aperture ratio increases in the previous study [31] in which an acoustic resonance frequency has been experimentally and analytically verified by assuming a one-dimensional sound field in a duct partitioned by a perforated plate, and similar results are obtained in this study. **Figure 33** shows the acoustic mode at each peak frequency in the cases of $\phi = 0$, 1, 2, and 4%. In the case of $\phi = 0\%$, the acoustic mode in the width direction strongly appears at 725 Hz. However, in the cases of $\phi = 1$, 2, and 4%, the acoustic mode in the duct width direction does not clearly appear, and the value of SPL decreases as the aperture ratio increases. Therefore, it was assumed that the perforated plate has an effect of attenuating sound.

5. Conclusions

1. The sound pressure level rises with an increase of the gap flow velocity by following the 5th power law when the gap velocity is low. However, when the gap velocity is high, the self-sustained tone is generated not by following the 5th power law.

2. Insertion of baffle plates in the tube bank decreases the natural frequency of the duct and increases the onset gap velocity of the self-sustained tone. Hence, the natural frequency of the duct does not seem to be related with the suppression of the self-sustained tone when the baffle plate is installed in the tube bank. The self-sustained tone is the most effectively suppressed by inserting the baffle in the entire tube bank because the baffle plate decreases the particle velocity and vorticity. Furthermore, there is a difference in the onset gap velocity of the self-sustained tone between when the baffle plate is inserted upstream and when it is inserted downstream.

3. The perforated plates installed on the duct walls suppress the self-sustained tone and increase the resonance frequency in the duct width direction. Consequently, if the perforated plates are installed on the duct walls, the self-sustained tone is assumed to be suppressed by an increase of the resonant frequency in the duct width direction and sound-absorbing effect of the perforated plates.

Author details

Masaaki Mori[1]* and Kunihiko Ishihara[2]*

1 Cybernet Systems CO., LTD, Tokyo, Japan

2 Department of Health and Welfare, Tokushima Bunri University, Japan

*Address all correspondence to: m-mori@cybernet.co.jp

References

[1] Baird RC. Pulsation-induced vibration in utility steam generation unit. Combustion. 1954;**25**:38-44

[2] Owen PR. Buffeting excitation of boiler tube vibration. Journal of Mechanical Engineering Science. 1965;7(4):431-439

[3] Chen YN. Flow-induced vibration and noise in tube-bank heat exchangers due to von karman streets. Journal of Engineering for Industry. 1968;**90**(1):134-146

[4] Fnunakawa M, Umakoshi R. The acoustic resonance in a tube bank. Bulletin of JSME. 1970;**13**(57):348-355

[5] Fitzpatric JA. The prediction of flow-induced noise in heat exchanger tube arrays. Journal of Sound and Vibration. 1985;**99**(3):425-435

[6] Blevins RD, Bresser MM. Acoustic resonance in heat exchanger tube bundles-part 1: Physical nature of the phenomena. Journal of Pressure Vessel Technology. 1987;**109**(3):275-281

[7] Weaver DS. Vortex shedding and acoustic resonance in heat exchanger tube arrays. Technology for the 90's. ASME. 1993:776-810

[8] Hamakawa H, Matsue H, Nishida E, Fukano T. Acoustic resonance and vortex shedding from tube banks of boiler plant. Journal of Fluid Science and Technology. 2008;**3**(6):805-813

[9] Hanson R, Mohany A, Ziada S. Effect of acoustic resonance on the dynamic lift of tube arrays. Journal of Fluids and Structures. 2009;**25**(3):80-94

[10] Ziada S, Oengören A. Vortex shedding in an in-line tube bundle with large tube spacings. Journal of Fluids and Structures. 1993;7(6):661-687

[11] Hamakawa H, Matsue H, Nishida E, Fukano T. Effect of arrangement of tube banks on acoustic resonance. Open Journal of Fluid Dynamics. 2012;**2**(4A):311-317

[12] Ishihara K, Tamehira T, Tsujii M, Ichimiya M. Study on a countermeasure of self-sustained tone by a baffle plate in boiler tube banks. Journal of Basic and Applied Physics. 2013;**2**(3):148-154

[13] Ishihara K, Nakaoka M. Study on a countermeasure using walls made of perforated plate for high level sound. In: Proceedings of ASME 2015 Pressure Vessels and Piping Conference Volume 4: Fluid-Structure Interaction

[14] Eisinger FL, Sullivan RE. Suppression of acoustic waves in steam generator and heat exchanger tube banks. Journal of Pressure Vessel Technology. 2003;**125**(2):221-227

[15] Feenstra PA, Weaver DS, Eisinger FL. Acoustic resonance in a staggered tube array: Tube response and the effect of baffles. Journal of Fluids and Structures. 2005;**21**(1):89-101

[16] Hamakawa H, Miyagi H, Nishida E, Fukano T. Effect of small cavities in in-line tube banks on vortex shedding. Transactions of the Japan Society of Mechanical Engineers, Part B. 2010;**76**(764):580-587

[17] Ishihara K, Takahashi T. Reduction of large noise generated in boiler and heat exchanger by flexible walls. Transactions of the Japan Society of Mechanical Engineers. 2013;**79**(802):1837

[18] Ingard U. On the theory and design of acoustic resonators. The Journal of the Acoustical Society of America. 1953;**25**:1037-1045

[19] Melling TH. The acoustic impedance of perforates at medium and high sound

pressure levels. Journal of Sound and Vibration. 1973;**29**(1):1-65

[20] Lawn C. The acoustic impedance of perforated plates under various flow conditions relating to combustion chamber liners. Applied Acoustics. 2016;**106**:144-154

[21] Hanson R, Mohany A, Ziada S. Flow-excited acoustic resonance of two side-by-side cylinders in cross-flow. Journal of Fluids and Structures. 2009;**25**:80-94

[22] Mohany A, Arthurs D, Boldu M, Hassan M, Ziada S. Numerical and experimental investigation of flow-acoustic resonance of side-by-side cylinders in a duct. Journal of Fluids and Structures. 2014;**48**:316-331

[23] Shahab K, Zaffar MK, Muhammad AM, Zafarullah K, Muhammad AJ, Mahmood AK, et al. Cross-Flow-Induced-Vibrations in Heat Exchanger Tube Bundles: A Review, Nuclear Power Plants. Europe: Intech; 2012. 4

[24] Mori M, Masumto T, Ishihara K. Numerical simulation of high level sound generated in boiler tube bank duct. International Journal of Engineering and Applied Sciences. 2018;**5**(1):19-24

[25] Mankbadi R, Lo S-C, Lyrintzis A, Golubev V, Dewan Y, Kurbatskii K. Hybrid LES-RANS simulations of a jet impinging on a flat plate. International Journal of Aeroacoustics. 2016;**15**(4-5):535-553

[26] Ishihara K, Tamehira T, Tsujii M, Ichimiya M. Study on countermeasure mechanism by baffle plate for self-sustained tone generated in boiler tube banks. Transactions of the Japan Society of Mechanical Engineers. 2012;**78**(787):749

[27] Howe MS. The dissipation of sound at an edge. Journal of Sound and Vibration. 1980;**70**(3):407-411

[28] Ishihara K. Study on a countermeasure for high level sound generated from boiler tube bank duct using walls made of perforated plate (Grasp critical aperture ratio and influence of cavity volume on suppression effect). Transactions of the Japan Society of Mechanical Engineers. 2017;**83**(848):456

[29] Mori M, Masumto T, Ishihara K. Numerical verification on a countermeasure for high level sound generated from boiler tube bank duct using walls made of perforated plate and cavity. International Journal of Science and Engineering. 2018;**7**(80):117-130

[30] Mori M, Masumto T, Ishihara K. Study on acoustic and flow induced noise characteristics of t-shaped pipe with square cross-section. Advances in Applied Acoustics. 2016;**5**:10-17

[31] Ishihara K, Kudo S, Masumoto T, Mori M. Study on acoustic natural frequency and its mode of one dimensional sound field partitioned by perforated plate. Transactions of the Japan Society of Mechanical Engineers. 2018;**84**(857):365

Chapter 5

Effective Low Frequency Noise Insulation Adopting Active Damping Approaches

Ming Yuan and Fan Yang

Abstract

In this chapter, effective low frequency noise insulation adopting active damping approaches are illustrated. In general, engineering plate structures suffer insufficient noise insulation performance in the low frequency band. To improve the structure's noise insulation performance, active damping methods can be utilized, which aim to suppress plate structure's efficient sound radiation modes. The collocated sensor/actuator configuration guarantees the control system's robustness, and simplifies the control law design. In the presented study, two control laws are proposed to add active damping to the structure. One control law is negative acceleration feedback (NAF) control and the other control law is filtered velocity feedback (FVF) control. The NAF control is suitable to control one specific mode and the FVF control is suitable to realize wide band vibration control. With respect to practical implementation, a carbon fiber reinforced plastic (CFRP) plate is served as the control target and active control laws are implemented on it. Experimental system for active control is presented in detail, and some practical advises are given to help readers to solve similar problems in a convenient way. The measured sound pressure and vibration results show effectiveness of the active damping treatment.

Keywords: noise insulation, low frequency, active damping, control laws, smart structure, real-time control

1. Introduction

Low frequency noise and vibration control has always been hot spot in academic and industry fields. For examples, in aircraft, vehicles, trains and industrial machines, large amplitude of structural responses are fluent in the low frequency ranges. These unwanted disturbances give rise to large decibels of noise and accelerating structure fatigue. Meanwhile, the noise's detrimental effect to human being's health has been well known [1].

In the mentioned examples above, plate structures are widely used. In general, the plate's structural parameters have been determined in advance, and lightweight structure is favorable for cost reduction. Accordingly, the plate structure usually have small thickness and low damping characteristics. When external noise excites the plate structure, the plate itself becomes an effective sound transmission path to the internal space [2, 3]. Specially, when the excitation frequency equals to the plate structure's natural frequency, large amplitude of vibration can impetus air particles strongly, making the structure itself can radiate noise efficiently [4].

To improve the plate's noise insulation performance, special treatments should be adopted to attenuate the sound transmission. An effective approach is adding damping to the plate. Accordingly, the structural vibration response is reduced, which can improve the structure's sound insulation performance substantially.

Among the damping approaches, passive damping is a simple and effective approach to realize vibration control. For instance, the viscoelastic material based approach has been applied widely in the engineering applications. However, this approach is only suitable to the mid-high frequencies and does not have satisfying performance in the low frequency range [5], especially when the installation have rigorous restrictions of added weight and spaces.

In another vein, active damping approaches have shown to be a good candidate to realize effective vibration control in the low frequency range. To perform active damping, an active control system is needed, which requires power supply. A classical active control system is composed by sensor, actuator, signal conditioner, controller and power amplifier [6].

In this study, we adopt piezoelectric smart material to realize structural sensing and actuating functions. The piezoelectric material can be integrated into the structure easily, which requires little mounting spaces [7]. The piezoelectric effect enables the structural signal being converted into electrical signal, and accelerometer is a commercialized product. The inverse piezoelectric effect enables the piezoelectric material being adopted as actuator, which follows the control signal well in the low frequency range. With proper design, it is anticipated that the plate's noise insulation performance can be improved significantly in the low frequency range after active control.

In Section 2, the general structural dynamic properties are characterized, and the interesting modes to be controlled are refined; in Section 3, the active control laws are proposed to increase the structure's damping behavior; in Section 4, some practical implementation issues for the active system are emphasized; in Section 5, a CFRP plate is selected as the control target and the performance of active sound insulation is evaluated.

2. Frequency band classification and interesting target modes

2.1 Frequency band classification

In general, the frequency range of the structural dynamics can be classified into three frequency bands, i.e., low frequency, mid frequency and high frequency [8, 9].

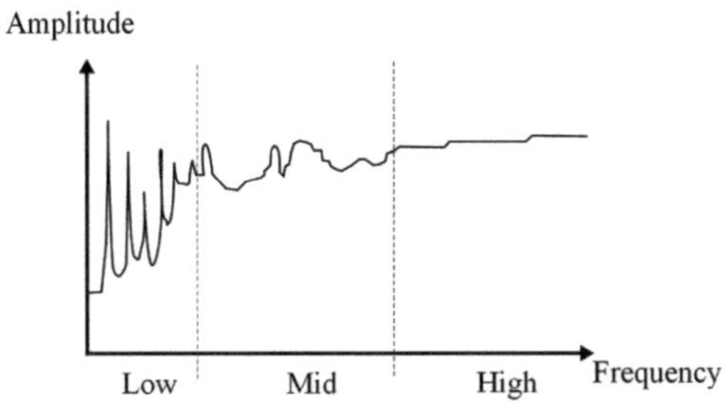

Figure 1.
Frequency band classification based on modal overlapping.

As shown in **Figure 1**, in the low frequency band, the structural responses are characterized by apparent vibration modes, peaks and troughs can be found at the corresponding frequency response plot. Besides, the structural uncertainties are small, making the structure's vibro-acoustic behavior can be predicted well with the help of finite element method.

When the frequency goes up into the mid and high frequency range, the modal overlapping increases and the peaks and troughs cannot be distinguished clearly from the frequency response plot. Meanwhile, the structural uncertainties and computation burden increase intensively, causing the finite element method being not suitable to handle such problem.

2.2 Interesting target modes

To realize effective active sound insulation, the vibration modes to be controlled must be selected carefully. Because the sound excitation is wideband in general, indicating multiple vibration modes can be excited at the same time, which brings difficulty to the control law design. Luckily, for each vibration mode, its sound radiation effectiveness differs significantly. If some modes contribute little to the far-field sound field, these modes can be omitted in the active control law design.

Besides, the mode selection has close relationship to the control loop's sensor/actuator placement. In general, the measured sensing signal should have sufficient signal to noise ratio (SNR) value. If the sensor/actuator is not placed properly, the control law's effectiveness and robustness will be decreased, owing to the low SNR value.

To place the sensor/actuator properly, and identify the interesting target modes, modal analysis can be adopted. Here, a rectangular plate is illustrated as an example and its first four mode shapes are presented in **Figure 2**.

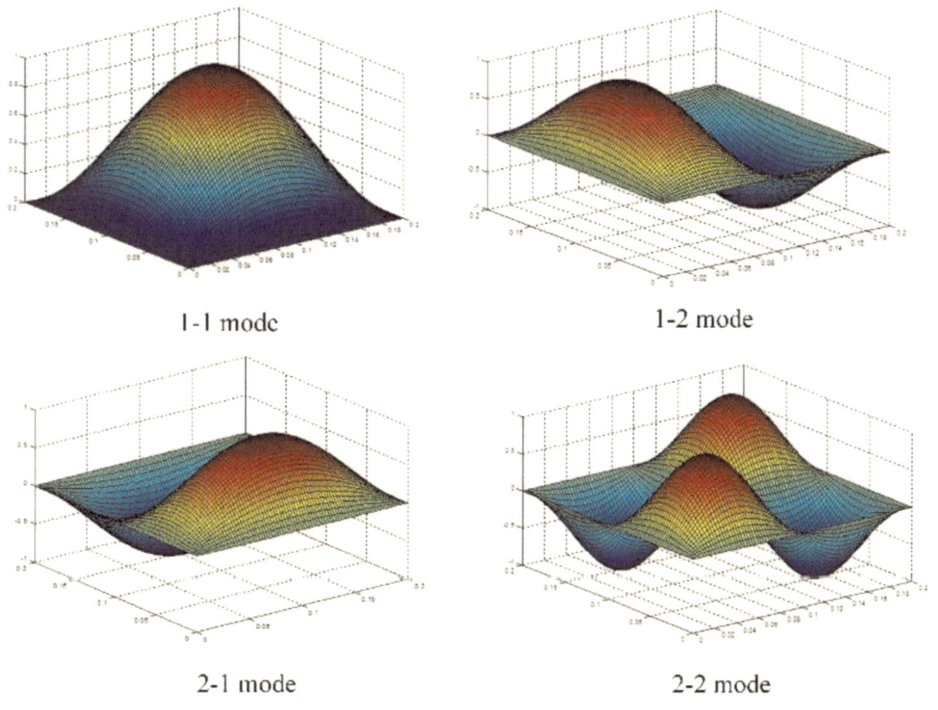

Figure 2.
Vibration mode shapes of rectangular plate.

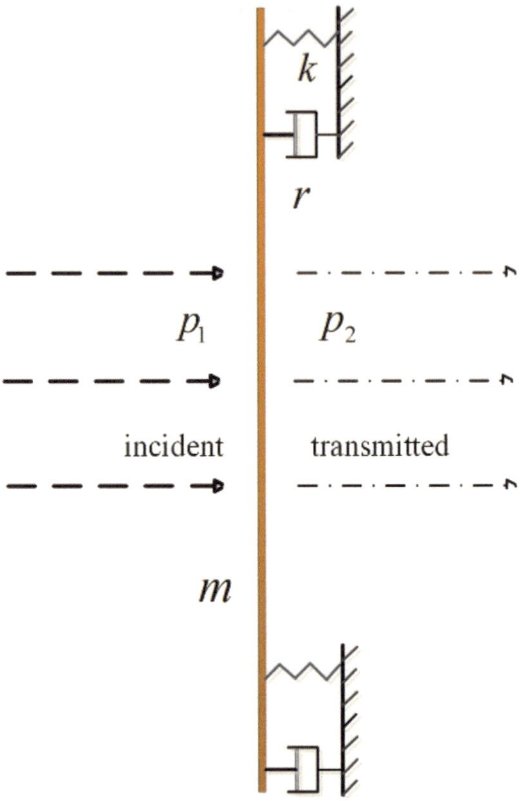

Figure 3.
Simplified model for noise insulation study.

According to these mode shapes, it is shown that the 2-2 mode (even-even mode) shape radiates like a quadrupole acoustic source. In the low frequency range, for such vibration mode, the neighboring push-pull phenomenon makes acoustic cancellation occurs strongly. Similarly, acoustic cancellation also occurs strongly at the 1-2 mode (odd-even mode) and 2-1 mode (even-odd mode), which reduce these modes' sound radiation effectiveness. The 1-1 mode (odd-odd mode) radiates like a monopole, which has the highest sound radiation ability.

Furthermore, the mode's radiation efficiencies can be quantized with different boundary conditions. For the simply supported boundary, the detailed equations can be found in Refs. [10, 11]. For the clamped boundary, the detailed equations can be found in Ref. [12].

These above analysis indicates that in the active sound insulation application, the main interested modes will be odd-odd modes. At the resonant frequency, and assuming the displacement is uniform, the structure can be simplified into a dynamic system with one degree of freedom.

As shown in **Figure 3**, the sound pressure from the incident side is p_1 and the sound pressure from the transmitted side is p_2. The structural displacement is η, damping coefficient is r, stiffness coeffcient is k, the dynamic system can be expressed as:

$$m\ddot{\eta} + r\dot{\eta} + k\eta = p_1 - p_2 \tag{1}$$

The volume velocity is assumed continuous at the plate's two sides, and the corresponding acoustic impedance is:

$$Z = (p_1 - p_2)/\dot{\eta} = j\omega m\left(1 - \frac{{\omega_0}^2}{\omega^2}\right) + r \tag{2}$$

where $\omega_0 = \sqrt{k/m}$ is the natural frequency.

According to the definition of sound transmission loss, when the incident sound wave is in vertical direction of the structure, the noise insulation metric can be calculated as:

$$\begin{aligned} R &= 10\log_{10}\left|\frac{p_1}{p_2}\right|^2 = 10\log_{10}\left|1 + \frac{Z}{2\rho_0 c_0}\right|^2 \\ &= 10\log_{10}\left|1 + \frac{j\omega m\left(1 - \frac{\omega_0^2}{\omega^2}\right) + r}{2\rho_0 c_0}\right|^2 \end{aligned} \tag{3}$$

where $\rho_0 c_0$ is the air's characteristic impedance.

The noise insulation suffers the lowest value at the resonant frequency, which can be expressed as:

$$R = 10\log_{10}\left|1 + \frac{r}{2\rho_0 c_0}\right|^2 \tag{4}$$

As shown in Eq. (4), to improve the noise insulation performance, the damping should be increased.

3. Active noise insulation control laws

3.1 Sensor and actuator placement

From the control engineering's perspective, sensor/actuator selection and placement correlate to the system's observability and controllability. For instance, for an interesting mode, if the sensor is placed at the structure's nodal point, no informative response can be obtained. The measured signal is buried in noise. Similarly, if the actuator is placed along the interesting mode's nodal line, this mode will be uncontrollable.

To improve the control law's robustness and facilitate control law design, the collocated sensor/actuator configuration is favorable. Under such configuration, the transmission path between the collocated actuator and sensor is the nearest, generates a minimum phase system. In other words, the magnitude and phase responses between the sensor and actuator is unique and the pole/zero appears in an interlacing way. Correspondingly, as shown in **Figure 4**, the resonant peaks and anti-resonant through will appear one by one. In the phase plot, 180° phase lag is generated across the resonant frequency and 180° phase lead is generated across the anti-resonant frequency. If the two adjacent modes are near, the modes' coupling makes the phase lag being smaller than 180°. In summary, the interlacing property makes the phase lag's variation always within 180°, which facilitates the control law design.

In practical implementation, the sensor and actuator's physical properties make the collocation in a limited frequency range. This means the phase lag between the sensor/actuator pair will beyond 180° above certain frequency limit. Therefore, to meet the Nyquist criterion [13], the controller's gain must be limited within a certain range.

3.2 Negative acceleration feedback (NAF) control

The NAF control law utilize the structural acceleration signal as the input signal. Then the control output is generated according to the NAF control law, which is written as:

$$H_{NAF}(j\omega) = g\frac{\omega_c^2}{-\omega^2 + 2\omega_c\zeta_{NAF}j\omega + \omega_c^2} \tag{5}$$

where g is the control gain, ω_c is the targeting control angular frequency, ζ_{NAF} is the controller's damping ratio parameter.

According to the analysis given in [14], at the targeting control angular frequency, the NAF control law generates active damping to the structure.

For a given control frequency, with different control damping parameters, the NAF controller's bode plot values is shown in **Figure 5**.

The damping ratio value should be paid attention for practical implementation. Clearly, if a lower damping ratio value is adopted, the control authority will be

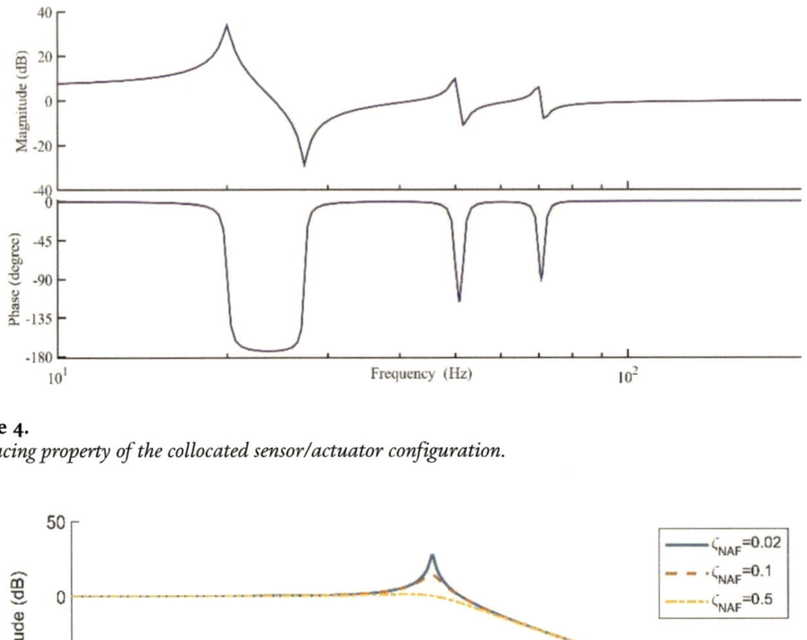

Figure 4.
Interlacing property of the collocated sensor/actuator configuration.

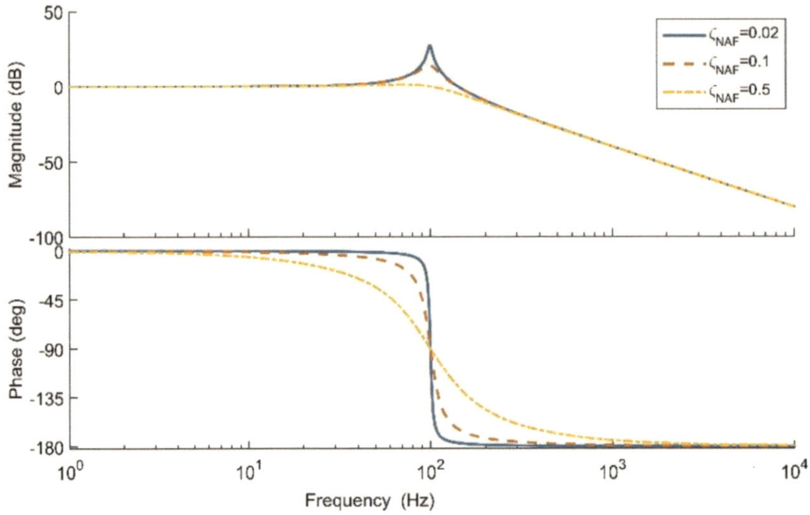

Figure 5.
Bode plot of the NAF controller with different control damping ratio parameters.

increased for a certain frequency. Because the controller can generate larger output at this frequency. However, the control performance will be reduced substantially when the target frequency is slightly biased from the realistic resonant frequency. If the damping value is too large, the control authority will be weak, and the added active damping effectiveness will be small.

The merit of NAF controller is: the control authority is exerted to one specific mode. The controller rolls-off at 40 dB/dec above the target control frequency. This roll-off characteristic can suppress high frequency noise effectively, which is favorable to guarantee the control gain.

Concerning multiple modes suppression, the same number of NAF controllers will be needed and the NAF controllers should be connected in parallel. However, the authors do not recommend using NAF control law to suppress more than three modes. One important reason is, when the interesting modes are adjacent, because the controller's output does not rolling off sufficiently, the control interferences are generated, making the control law's tuning process being complicated.

3.3 Filtered velocity feedback (FVF) control

Direct velocity feedback (DVF) control utilizes the structural velocity as the feedback signal directly. The controller is a constant gain, which can generate active damping to the structure [15].

Because the DVF control law does not provide roll-off property, the control authority is limited for practical experimental test. However, in the sound insulation application, to realize global reduction, wideband active damping is required essentially.

To fulfill this task, the DVF control law must be modified and Filtered Velocity Feedback (FVF) control is proposed to improve the control performance. The FVF control law behaves like an electrical dynamic absorber, which can be deduced according to Ref. [16].

In the frequency domain, the FVF controller can be expressed as:

$$H_{FVF}(j\omega) = g\frac{j\omega + 2\omega_n\zeta_{FVF}}{-\omega^2 + 2\omega_n\zeta_{FVF}j\omega + \omega_n^2} \tag{6}$$

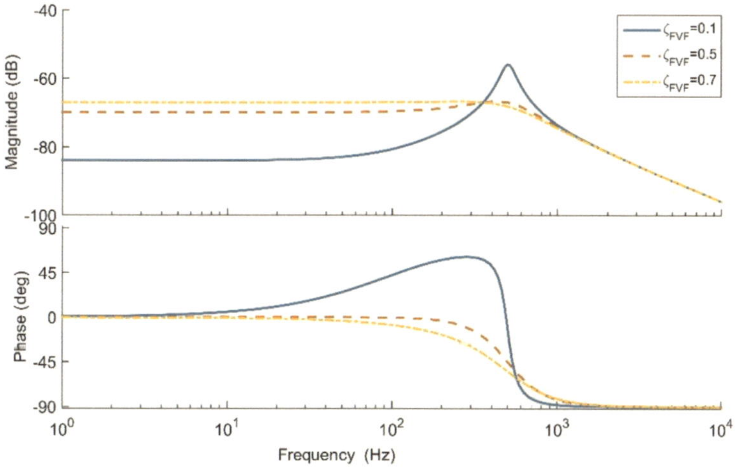

Figure 6.
Bode plot of the FVF controller with different control damping ratio parameters.

where g is the control gain, ω_n is the controller's upper frequency, ζ_{FVF} is the controller's damping ratio.

For a given control frequency, the FVF controller's Bode plot with different control damping values is shown in **Figure 6**.

As shown in **Figure 6**, in the low frequency, the controllers behaves like a DVF controller but the control output rolls off at 20 dB/dec after the controller's upper frequency. Therefore, active damping is generated in the low frequency range.

Unlike the NAF controller, the damping ratio of FVF controller should be high enough to generate sufficient gain below the upper frequency. For instance, $\zeta_{FVF} = 0.7$ is a good choice.

If acceleration signal is used as the control input instead of velocity signal, the modified FVF control law becomes into:

$$H'_{FVF}(j\omega) = g\frac{j\omega + 2\omega_n\zeta_{FVF}}{j\omega\left(-\omega^2 + 2\omega_n\zeta_{FVF}j\omega + \omega_n^2\right)} \tag{7}$$

4. Implementation considerations

4.1 Real-time control platform

To implement the control law, a real-time operating system (RTOS) is required to maintain the control law operating in a deterministic way. Normally, the sensing signal is sampled from the analog input channel, and the data acquisition device is switched to hardware-timed single point mode. The real-time control system guarantees the control law being updated within the period of two consecutive sampling points. After the control algorithm is updated, the output signal is sent to the analog output channel.

In general, the control law's thread should have the highest priority during the execution. This mechanism makes the control law can be implemented in a deterministic way. The other functions of the RTOS, such as signal monitoring, data logging and network communications are usually assigned to low execution priorities.

For rapid control system prototyping, some examples of the real-time target are: PXI target [17], CompactRIO target [18], Speedgoat target [19] and dSPACE target [20]. These real-time platforms' corresponding photo is shown in **Figure** 7.

4.2 Delay minimization

With respect to the feedback control system, time delay is highly unwanted. However, inside the feedback control system, each component can introduce some kind of delay, which generates adverse influence to system's stability and performance.

For a pure delay system, the phase lag (in degree) can be expressed as:

$$\varphi = -57.3 \times \omega \times T \tag{8}$$

where ω is the angular frequency and T is the time caused by system delay.

If the sensor and actuator are placed in the collocated configuration, the physical signal's propagation delay from the actuator to the sensor is minimized. Therefore, the digital control system becomes the primary time delay source, which has close relationship with the sampling rate.

To minimize the delay caused by data sampling, the sampling rate should be high enough. In general, the sample rate should be at least 10 times higher than the

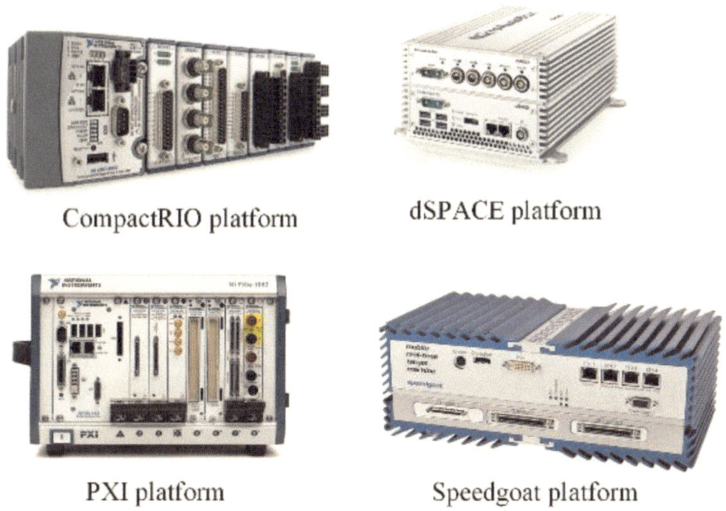

Figure 7.
Real-time control platforms for rapid prototyping.

interesting bandwidth's upper frequency. However, the achieved sampling rate is also restricted by the controller. When the control law requires intensive computation effort and the controller is not fast enough, the control law will not be finished in time. In such circumstance, to meet the deterministic mechanism, the sampling rate should be reduced.

In summary, to determine the control law's updating rate, the control bandwidth, computation burden and real-time target's performance should be taken into consideration simultaneously.

4.3 Filter

Filters are widely used in active control systems, which can suppress excessive noise signal. Crucially, if the control law itself does not have the roll-off characteristic in the high frequency, low pass filter will be always favorable. The high pass filter can also be utilized to suppress the ultra-low frequency noise.

The filter's cut-off frequency and order should be determined with respect to the control plant. For the realization form, analog filter is recommended. Because analog filter has much less propagation delay than the digital filter.

5. Case studies

As shown in **Figure 8**, the sound insulation test was performed in an anechoic room. The control target is a stiffened CFRP plate and the dimension parameter is 840 mm × 840 mm × 1 mm. The plate is mounted on an aluminum cavity, with a thickness of 16 mm. A loudspeaker is placed inside the cavity, which generates acoustic excitation the stiffened plate. The CFRP plate is clamped to the aluminum cavity using screws. The edges between the CFRP plate and aluminum box are sealed to prohibit air leakage.

According to the sound insulation mass law, because the box's thickness is much larger than the CFRP plate, the sound transmission path from the internal cavity to

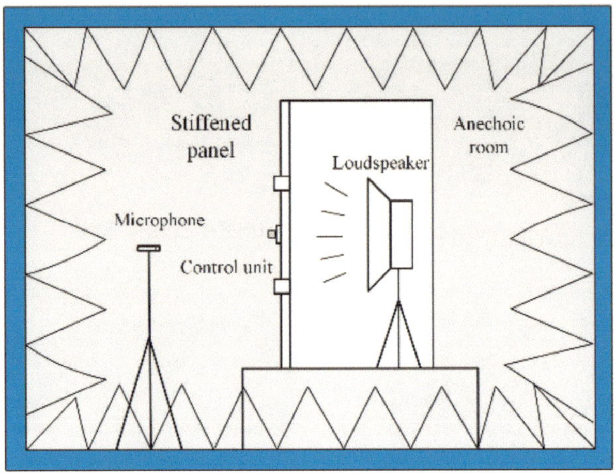

Figure 8.
Schematic diagram for the active sound insulation test.

Figure 9.
Photograph of the test structure.

the outside is the latter. To evaluate the control law's performance, in front of the CFRP plate, a microphone sensor (G.R.A.S 40 ph) is used to monitor the sound pressure level (SPL) value. The distance between the microphone sensor and CFRP plate is 1 m.

For the control system implementation, three independent control channels are adopted. Correspondingly, as shown in **Figure 9**, three piezoelectric PZT-5H patches are served as actuators, which are labeled as 1, 2, 3.

The detailed components of control system are shown in **Figure 10**. Collocated accelerometers are bonded to the central of the piezoelectric patches, which are used for structural sensing. Control algorithms are programed using LabVIEW™ and subsequently compiled and downloaded into a CompactRIO target for real-time implementation. A Virtex-5 LX110 FPGA chip guarantees the control loops can be implemented with high throughput and in parallel physically [21].

The digital control loop's updating rate is set to 20 kHz. To transform the continuous transfer function into discretized form, bilinear transformation method is adopted. The equation of the bilinear transformation is shown as follows:

$$z = \frac{1 + Ts/2}{1 - Ts/2} \tag{9}$$

where T is the sampling time.

The components of the control system is shown in **Figure 10**. Band-limited noise (20–500 Hz) is sent to the loudspeaker, which can excite multiple vibration modes of the CFRP panel. Besides, the 1-1 mode is a global vibration mode, which needs more control effort. In view of the above circumstances, a hybrid control law is proposed to solve the problem.

In the hybrid scheme, the NAF control puts control authority to the plate's 1-1 mode and FVF control puts control authority to multiple modes of the plate.

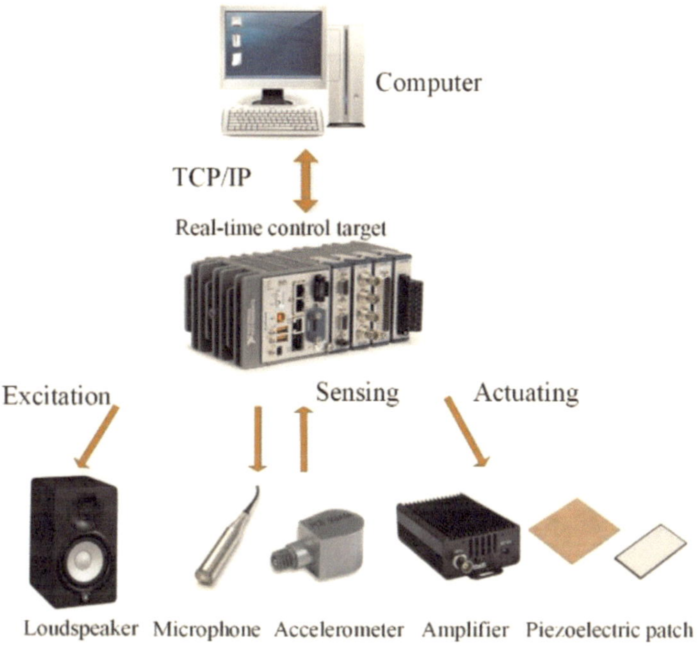

Figure 10.
Components of the control system.

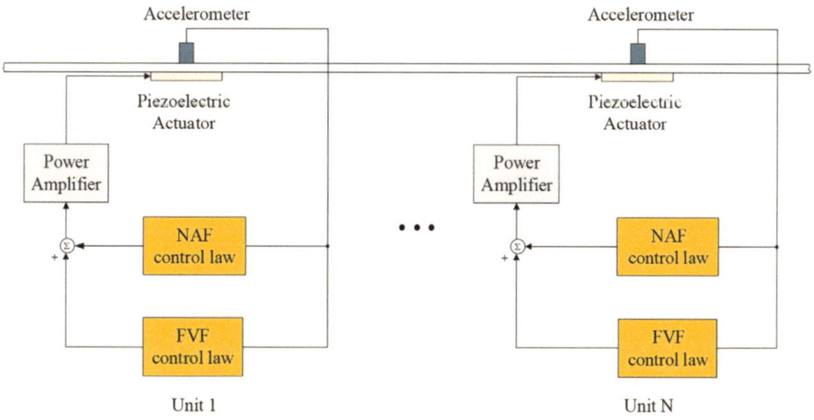

Figure 11.
Schematic diagram of the hybrid control system.

Therefore, it is anticipated that this hybrid control law can exert sufficient active damping to the CFRP panel, which is suitable for the noise insulation application.

The schematic diagram of the hybrid control system is shown **Figure 11**.

When the active insulation system starts to work, the acceleration signal spectrums before control and after control are shown in **Figure 12**.

From the measured data, the proposed hybrid control scheme can realize wideband vibration reduction. For instance, at 83 Hz, the vibration reduction value is 13 dB; at 143 Hz, the vibration reduction value is 6.6 dB; at 194 Hz, the vibration reduction value is 5.3 dB; at 216 Hz, the vibration reduction value is 5.2 dB; at 273 Hz, the vibration reduction value is 4.3 dB; at 305 Hz, the vibration reduction value is 5.9 dB.

The SPL spectrums before control and after control are shown in **Figure 13**.

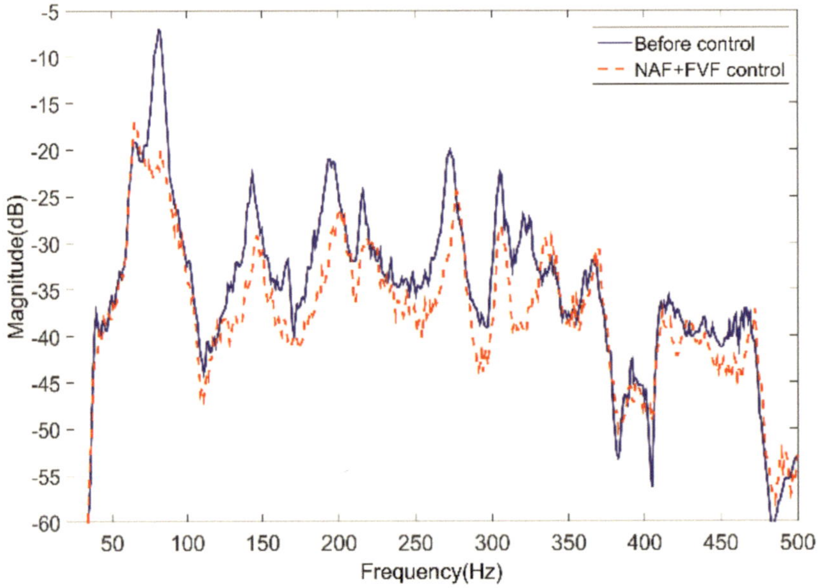

Figure 12.
Measured acceleration signal (solid line: before control, dashed line: after control).

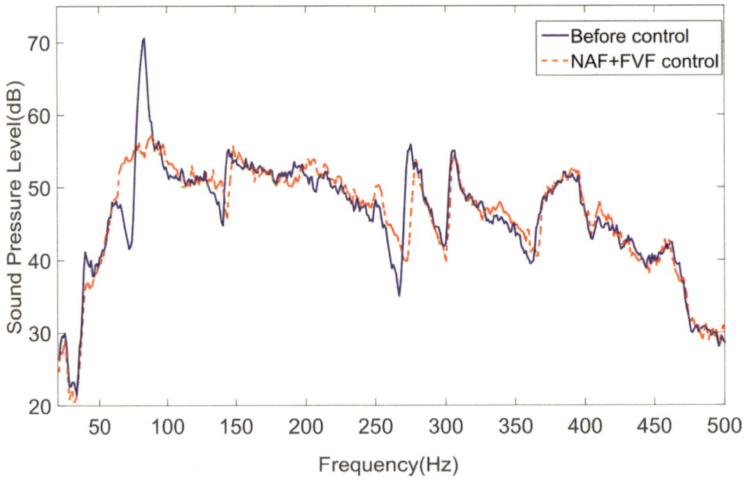

Figure 13.
Measured sound pressure level (solid line: without control, dashed line: with control).

At 83 Hz, as anticipated, this first vibration mode contribute to the SPL spectrum significantly. After the active damping treatment, at this frequency, the SPL reduction value is up to 15.3 dB. For the entire interesting frequency band, the SPL reduction value comes up to 7.3 dB. These measured experimental results prove that the proposed active damping control method can improve the structure's sound insulation performance in an effective way.

6. Conclusions

In this chapter, the active damping approaches are adopted to improve structure's sound insulation performance in the low frequency range. The collocated sensor/actuator configuration is adopted to simplify the control law design. The NAF control law is proposed to suppress the plate's 1-1 mode, which radiates sound effectively and needs more control authority. The FVF control is proposed to suppress multiple vibration modes in wide range. In the experimental study, a CFRP panel is utilized as the control target and the NAF and FVF control laws are combined together to generate active damping to the CFRP structure. Experimental test results show the hybrid control law can realize wideband active damping and improve the plate's sound insulation performance significantly.

Acknowledgements

This work was funded by the National Natural Science Foundation of China (grant numbers 61701250, 51605202), the Natural Science Foundation of Jiangsu Province (grant number BK20160895, BK20160550).

Conflict of interest

The authors declare no conflict of interest.

Author details

Ming Yuan[1*] and Fan Yang[2*]

1 School of Automation, Nanjing University of Posts and Telecommunications, Nanjing, China

2 School of Naval Architecture and Ocean Engineering, Jiangsu University of Science and Technology, Zhenjiang, China

*Address all correspondence to: yuanming@njupt.edu.cn and yangf@just.edu.cn

References

[1] Ising H, Kruppa B. Health effects caused by noise: Evidence in the literature from the past 25 years. Noise & Health [Internet]. 2004;**6**(22):5-13. Available from: http://www.noiseandhealth.org/article.asp?issn=1463-1741

[2] Wilby JF. Aircraft interior noise. Journal of Sound and Vibration. 1996; **190**(3):545-564

[3] Yuan M, Ohayon R, Qiu J. Decentralized active control of turbulent boundary induced noise and vibration: A numerical investigation. Journal of Vibration and Control [Internet]. 2016;**22**(18):3821-3839. DOI: 10.1177/1077546314566441

[4] Fahy FJ, Gardonio P. Sound and Structural Vibration: Radiation, Transmission and Response. Oxford: Academic Press; 2007

[5] Baz AM. Active and Passive Vibration Damping. 1st ed. Hoboken, NJ, USA: Wiley; 2019. pp. 1-5

[6] Yuan M, Ji H, Qiu J, Ma T. Active control of sound transmission through a stiffened panel using a hybrid control strategy. Journal of Intelligent Material Systems and Structures [Internet]. 2012;**23**(7):791-803. DOI: 10.1177/1045389X12439638

[7] Wang Y, Inman DJ. A survey of control strategies for simultaneous vibration suppression and energy harvesting *via* piezoceramics. Journal of Intelligent Material Systems and Structures. 2012;**23**(18):2021-2037

[8] Morand HJ-PP, Ohayon R. Fluid Structure Interaction. New York: John Wiley; 1995. 212 p

[9] Ohayon R, Soize C. Advanced Computational Vibroacoustics: Reduced-Order Models and Uncertainty Quantification. New York: Cambridge University Press; 2014. pp. 1-127

[10] Wallace CE. Radiation resistance of a rectangular panel. The Journal of the Acoustical Society of America [Internet]. 1972;**51**(3B):946-952. DOI: 10.1121/1.1912943

[11] Pierre RLS Jr, Koopmann GH, Chen W. Volume velocity control of sound transmission through composite panels. Journal of Sound and Vibration. 1998; **210**(4):441-460

[12] Berry A, Guyader J, Nicolas J. A general formulation for the sound radiation from rectangular, baffled plates with arbitrary boundary conditions. The Journal of the Acoustical Society of America. 1990;**88**(6): 2792-2802

[13] Franklin GF, Powell JD, Naeini AE. Feedback Control of Dynamic Systems. 7th ed. Upper Saddle River, New Jersey: Pearson Press; 2015

[14] Yuan M, Qiu J, Ji H, Zhou W, Ohayon R. Active control of sound transmission using a hybrid/blind decentralized control approach. Journal of Vibration and Control [Internet]. 2013;**21**(13):2661-2684. DOI: 10.1177/1077546313514758

[15] Balas MJ. Direct velocity feedback control of large space structures. Journal of Guidance, Control, and Dynamics [Internet]. 1979;**2**(3):252-253. DOI: 10.2514/3.55869

[16] Yuan M. Compact and efficient active vibro-acoustic control of a smart plate structure. International Journal of Engineering. 2016;**29**(8):1068-1074

[17] PXI Systems [Internet]. Available from: http://www.ni.com/en-us/shop/pxi.html

[18] CompactRIO Systems [Internet]. Available from: http://www.ni.com/en-us/shop/compactrio.html

[19] Speedgoat Systems [Internet]. Available from: https://www.speedgoat.com/

[20] dSPACE Systems [Internet]. Available from: https://www.dspace.com/shared/data/pdf/2018/dSPACE_Systems.pdf

[21] Yuan M. Field programmable gate Array implementation of active control laws for multi-mode vibration damping. International Journal of Engineering-Transactions B: Applications. 2016; **29**(2):229-235

Chapter 6

Analysis of Energy Relations between Noise and Vibration Produced by a Low-Field MRI Device

Jiří Přibil, Anna Přibilová and Ivan Frollo

Abstract

Magnetic resonance imaging (MRI) tomography is often used for noninvasive scanning of various parts of a human body without undesirable effects present in X-ray computed tomography. In MRI devices, slices of a tested subject are selected in 3D coordinates by a system of gradient coils. The current flowing through these coils changes rapidly, which results in mechanical vibration. This vibration is significant also in the equipment working with a low magnetic field, and it causes image blurring of thin layer samples and acoustic noise significantly degrading a speech signal recorded simultaneously during MR scanning of the vocal tract. There are always negative physiological and psychological effects on a person exposed to vibration and acoustic noise. In order to minimize these negative impacts depending on intensity and time duration of exposition, we mapped relationship between energy of vibration and noise signals measured in the MRI scanning area and its vicinity.

Keywords: magnetic resonance imaging, acoustic noise, mechanical vibration, statistical analysis, low magnetic field environment

1. Introduction

The magnetic resonance imaging (MRI) method is successfully used for monitoring progress in therapy after vocal fold cancer surgery or for monitoring of the implanted cartilage in legs or arms, and/or the process of regeneration in different tissues, etc. In the case of the open-air MRI device, a weak magnetic field (up to 0.2 T) is usually generated by a pair of permanent magnets. Between these magnets, the gradient system consisting of 2×3 planar coils is situated together with the RF receiving/transmitting coils surrounding the tested object [1]. Slices of a tested object are selected in 3D coordinates by a gradient system consisting of planar coils parallel to the magnets. A rapidly changing current flowing through the gradient coils produces significant mechanical vibration [2, 3] causing blurring of images of thin layer samples and acoustic noise significantly degrading the speech signal recorded simultaneously during MR scanning of the human vocal tract [4, 5]. Acoustic noise has always negative physiological and psychological consequences on the exposed person depending on the noise intensity and time duration of noise exposure [6]. In order to minimize these negative factors, this work is focused on mapping of energy relationship between vibration and noise signals measured in the MRI scanning area and its vicinity with the final aim to choose the proper scan sequence and its parameters—repetition time (TR), echo time (TE), orientation of scan slices, etc. Apart from real-time recording of the vibration and noise signals, the sound pressure level (SPL) was measured by a sound level meter using frequency weighting to match human perception of noise. The measured data and recorded signals were further processed off-line—the determined energetic features were statistically analyzed and the results were compared visually and numerically.

2. Subject and methods

As mentioned above, the open-air MRI device is primarily used in medical diagnostics, so designation of three planes formed by x, y, and z axes follows medical terminology used for human body planes [7]. The plane dividing the body vertically into ventral (anterior) and dorsal (posterior) parts is called a coronal (frontal) plane. The second vertical plane dividing the body to left and right sides is a sagittal plane. The horizontal plane that divides the human body into superior (upper) and inferior (lower) parts is called a transverse (cross-sectional) plane. During sequence execution, the gradient coil pair corresponding to the chosen scan orientation is activated, it consequently vibrates, and acoustic noise is radiated in the surrounding air. Two basic types of sequences called spin echo (SE) and gradient echo (GE) arising from MRI physical principles [8] are preferred in this type of MRI device. The volume size of the tested object/subject is another important factor having an influence on the intensity of the produced vibration and noise in the scanning area of the MRI device. A tested person/sample/phantom as a part of the whole vibrating mechanical system changes the overall mass, stiffness, and damping by loading the lower gradient coil structure in the patient's bed.

2.1 Sensors for measurement in a weak magnetic field environment

If the vibration and noise signals are recorded during MR scanning, interaction with the stationary magnetic field B_0 in the scanning area must be eliminated; otherwise, the quality of the acquired images would not be preserved. It means that the vibration sensors placed in the MRI scanning area with the static magnetic field cannot contain any part made from a ferromagnetic material. In MRI equipment, working with a weak magnetic field the interaction problem can be solved by a proper choice of the measuring device and its arrangement. Usually, it is sufficient to locate it in an adequate distance from the noise signal source outside the magnetic field area. Since the noise intensity as well as its spectral properties depends on the position of the measuring instrument, the recording/measuring microphone must have high sensitivity, an appropriate pickup pattern, type of the microphone, and a position in regard to the central point of the MRI scanning area (distance, direction angle, working height). The best solution is to use a microphone with a variable pattern having two diaphragms that share a common back plate. Such a microphone behaves as two back-to-back cardioid microphones. If one membrane is connected to a constant polarization voltage and the second one is polarized by a variable voltage, principally any directional pattern can be created. Basic omnidirectional, figure-of-eight, and cardioid patterns corresponding to both same voltages of the same polarity, the opposite polarity, and one zero voltage are represented in an ideal form by a polar equation:

$$\rho(\theta) = A + B \cdot \cos\theta, \qquad A + B = 1, \tag{1}$$

where $A = 1$, $B = 0$ for omnidirectional, $A = 0$, $B = 1$ for figure-of-eight, and $A = 0.5$, $B = 0.5$ for cardioid directional patterns.

The noise distribution in the scanning area of the MRI equipment and its neighborhood has to be mapped prior to the selection of the proper recording microphone location. C-weighting was used for SPL measurement to accommodate the objective noise intensity to the subjective loudness at high sound levels. The C-weighting filter frequency response in s-domain is given by the equation

$$H(s) = G \cdot \frac{(2\pi f_2)^2 \cdot s^4}{(s + 2\pi f_1)^2 \cdot (s + 2\pi f_2)^2}, \tag{2}$$

where $f_1 = 20.6$ Hz, $f_2 = 12{,}194$ Hz, and $20 \log G = 0.062$ dB [9]. To get the transfer function of the digital IIR filter, the frequency scale is warped by the bilinear transform from s-plane to z-plane

$$s \rightarrow 2 \cdot \frac{1 - z^{-1}}{1 + z^{-1}}. \tag{3}$$

The sensors measuring vibration signals are placed inside the MRI scanning area where the basic stationary magnetic field of the MRI device is present together with the superimposed pulse magnetic field generated by the gradient system as well as the high voltage field originated during activation of the excitation RF coil. These fields would disturb a signal picked up by the sensor from ferromagnetic material or damage electronics integrated with the sensor [10, 11], which can be avoided using the vibration sensor with a piezoelectric transducer. The sensor must have good sensitivity and maximally flat frequency response with the frequency range covering the vibration and noise harmonic frequencies that fall into the low band due to frequency-limited gradient pulses. As a similar frequency range can be found in basic processing of speech signals, it is very important in the case of 3D scanning of the human vocal tract by MRI with parallel recording of a speech signal [5].

The mentioned requirements imposed on the vibration sensor can be met by the sensor for acoustic musical instruments [12]. Its first usage in the magnetic field environment must be preceded by a calibration procedure and a measurement of its sensitivity and frequency response. The measured frequency response is used to determine a correction curve for filtering of the picked-up vibration signal and consecutive linearization operation that has effect on correctness of all analyzed spectral properties determined from the vibration signals—see the block diagram in **Figure 1**. The correction filter is proposed by a standard procedure of second-order shelving filter design [13]:

$$H(z) = \frac{b_0 + b_1 z^{-1} + b_2 z^{-2}}{a_0 + a_1 z^{-1} + a_2 z^{-2}}. \tag{4}$$

For the sampling frequency f_s, the polynomial filter coefficients $a_{0,1,2}$ and $b_{0,1,2}$ are derived from three input parameters—gain G, mid-point frequency f_c, and quality factor Q—in the following manner:

$$A = 10^{\frac{G}{20}}, \omega_c = 2\pi \cdot \frac{f_c}{f_s}, \tag{5}$$

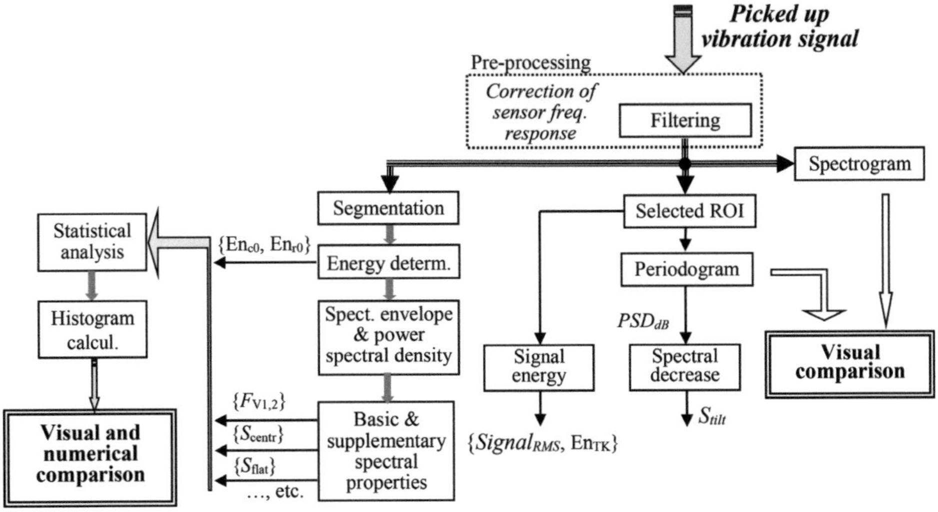

Figure 1.
Block diagram of processing of the picked-up vibration signal.

$$q_s = \frac{\sqrt{A}}{Q} \cdot \sin \omega_c, \; q_{c1} = (A+1) \cdot \cos \omega_c, \; q_{c2} = (A-1) \cdot \cos \omega_c, \tag{6}$$

$$a_0 = A + 1 - q_{c2} + q_s, \; a_1 = 2 \cdot [A - 1 - q_{c1}], \; a_2 = a_0 - 2q_s, \tag{7}$$

$$b_0 = A \cdot [A + 1 + q_{c2} + q_s], \; b_1 = -2A \cdot [A - 1 + q_{c1}], \; b_2 = b_0 - 2q_s. \tag{8}$$

2.2 Features for description of vibration and noise signals

Several methods can be used to determine the energy of a periodical signal:

- The standard root mean square (RMS) is calculated from a signal $x(n)$ in a defined region of interest (ROI) with the length of M samples

$$Signal_{RMS} = \sqrt{\frac{1}{M} \sum_{n=1}^{M} |x(n)|^2}. \tag{9}$$

- The absolute value of the mean of the Teager-Kaiser energy operator O_{TK} [14] is used to calculate the energy En_{TK}

$$O_{TK} = x(n)^2 - x(n-1) \cdot x(n+1), \qquad En_{TK} = abs\left(\frac{1}{M-2} \sum_{n=1}^{M-2} O_{TK}(n)\right). \tag{10}$$

- The frame energy is estimated by the first cepstral coefficient c_0 or the autocorrelation coefficient r_0 after processing the signal $x(n)$ in frames using N_{FFT}-point FFT to compute magnitude spectrum and power spectrum $|S(k)|^2$,

$$En_{c0} = \sqrt{\left[\prod_{k=1}^{N_{FFT}/2} |S(k)|^2\right]^{\frac{2}{N_{FFT}}}}, \qquad En_{r0} = \frac{2}{N_{FFT}} \sum_{k=1}^{N_{FFT}/2} |S(k)|^2. \tag{11}$$

For basic visual comparison of spectral properties of the recorded vibration and noise signals, the periodogram representing an estimate of the power spectral density (PSD) can be successfully used. The basic spectral properties can be determined from the spectral envelope, and subsequently, the histograms of spectral values can be calculated and compared. They also include the basic resonance frequencies F_{V1} and F_{V2} and their ratios, and the spectral decrease (tilt-S_{tilt}) as the degree of fall of the power spectrum calculated by a linear regression using the mean square method.

The supplementary spectral features describe the shape of the power spectrum of the noise signal. The spectral centroid (S_{centr}) determines a center of gravity of the spectrum—the average frequency weighted by the values of the normalized energy of each frequency component in the spectrum

$$S_{centr} = \frac{f_s}{N_{FFT}} \cdot \frac{\sum_{k=1}^{N_{FFT}} k|S(k)|^2}{\sum_{k=1}^{N_{FFT}} |S(k)|^2}. \tag{12}$$

The spectral flatness (S_{flat}) is useful to determine the degree of periodicity in the signal, and it can be calculated as a ratio of the geometric and the arithmetic mean values of the power spectrum

$$S_{flat} = \frac{\left[\prod_{k=1}^{N_{FFT}} |S(k)|^2\right]^{\frac{2}{N_{FFT}}}}{\frac{2}{N_{FFT}} \sum_{k=1}^{N_{FFT}} |S(k)|^2}. \tag{13}$$

The spectral entropy is a measure of spectral distribution. It quantifies a degree of randomness of spectral probability density represented by normalized frequency components of the spectrum. The Shannon spectral entropy (SHE) can be calculated using the following formulas:

$$SHE = -\sum_{k=1}^{N_{FFT}} |S(k)|^2 \log_2 |S(k)|^2. \tag{14}$$

3. Description of performed measurements and experiments

The performed measurements were focused on analysis of vibration and noise conditions in the scanning area and in the neighborhood of the open-air MRI equipment E-scan Opera by Esaote S.p.A., Genoa [15] located at the Institute of Measurement Science, SAS, Bratislava. The experiments were realized in four steps: in the preliminary phase, the calibration was carried out, and the sensitivity and the frequency response of the used vibration sensor were determined. Next, the noise was measured using different directional patterns of the pickup microphone and the influence of the pickup pattern on the spectral properties of the recorded noise signal was analyzed. Then, the main vibration and noise measurement and recording experiment were realized. The recorded signals were subsequently processed and statistically analyzed. Finally, a detailed analysis of the influence of chosen scan parameters on the time duration of the used MR sequences and on the quality factor of the MR images was performed with the aim to find a suitable setting to minimize exposition of the examined persons to noise and vibration.

3.1 Calibration of vibration sensors suitable for measurement in the low magnetic field environment

The calibration and measurement experiments were realized with the help of the main devices: the Audio Precision System One including two programmable input and output channels for simultaneous measurement of electrical signals from the vibration sensors mounted on the Vibration Exciter ESE 201 located at the Institute of Electronics and Photonics, FEE&IT SUT, Bratislava. As a reference sensor, the accelerometer KD35a from the company Metra Mess- und Frequenztechnik was used. The sensor sensitivity of this standardized accelerometer is guaranteed, and it operates over a frequency range from 50 Hz to 10 kHz. Three types of vibration sensors having good response in the lower audio frequency range up to 2 kHz were tested within this work:

- Cejpek SB-1 with the thin circular brass disc of 0.25-mm thickness and 27.5-mm diameter designed primarily for pickup of a musical sound of a contrabass (further called as "*SB-1*"),

- Shadow SH-SB2 double bass pickup with two disc transducers of 0.5-mm thickness and 22.5-mm diameter (further called as "*SB2a,b*"),

- RFT heart microphone device HM 692 comprising a piezo-electric element integrated in the 1-mm thin aluminum metal cover with 30-mm diameter (further called as "*HM692*").

The sensors were mounted on the plate of the vibration exciter as shown in the detailed photo of the arrangement of the sensors in the right part of **Figure 2**. The output voltage for supply of this exciter and the signal from the calibrated sensors were checked parallel by the digital oscilloscope Rigol DS1102E. Two types of the parameters of the vibration sensors were measured and compared in our experiment:

- relative sensitivity at the reference frequency f_{ref} = 125 Hz,

- frequency response in the range from 20 Hz to 2 kHz at the chosen output voltage of the vibration exciter (U_{excBa0} = 360 mV).

Dependence of the sensor's sensitivity on the excitation voltage for all three sensors is presented in **Figure 3a**. The reference voltage sensitivity B_{a0} of the SB-1

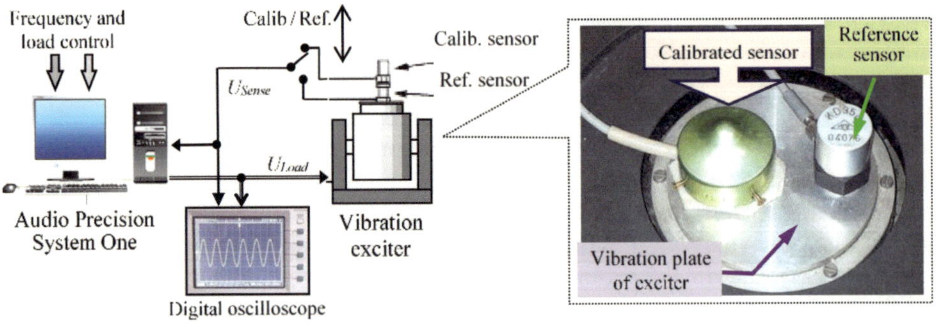

Figure 2.
Principle block diagram of the used calibration and measurement method together with a detailed photo of practical mounting of the sensors on the plate of the vibration exciter.

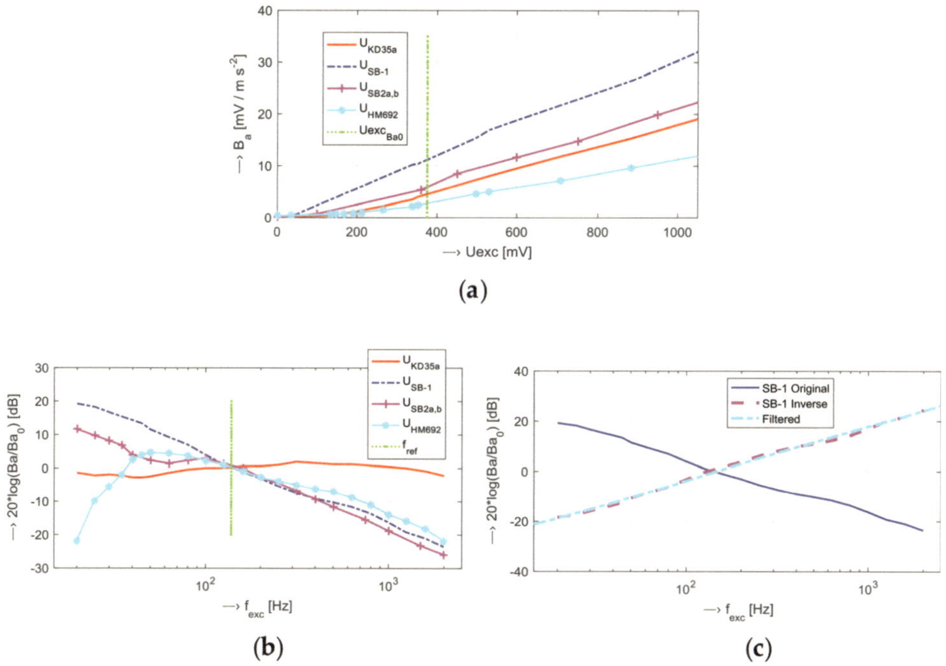

Figure 3.
Graph of: (a) measured sensors' sensitivities, (b) frequency responses in the range 20 Hz to 2 kHz measured and recalculated in [dB], and (c) correction frequency response for the SB-1 sensor linearization using the shelving filter (b): f_{ref} *= 125 Hz,* $Uexc_{Bao}$ *= 360 mV,* B_{ao} *= {3.69 (KD35a), 12.9 (SB-1), 5.65 (SB2ab), and 2.45 (HM692)} mV/m* s^{-2}*.*

sensor was determined from this graph. Comparison in **Figure 3b** shows that the measured frequency responses of SB-1, SB2ab, and HM692 are rotated by a slope of about −20 dB per decade with respect to the frequency response of KD35a. As the reference KD35a is an acceleration sensor, it emerges that the remaining three sensors are velocity ones. The calculated inverse frequency response of the SB-1 is drawn by the magenta dashed line together with the correction frequency response obtained by shelving equalization that is plotted by the cyan dot-dash line in **Figure 3c**. The effect of this shelving filter on the time-domain vibration signal, its frequency-domain periodogram with chosen spectral features, and the spectrogram can be seen in **Figure 4**.

3.2 Analysis of the influence of the directional pattern of the pickup microphone on the spectral properties of the recorded noise signal

Acoustic noise measurement in the MRI neighborhood was realized in the directions of 30, 90, and 150°, at the distance of 60 cm from the central point of the scanning area, and at the height of 85 cm from the floor—see the principal arrangement photo in **Figure 5**. In this noise recording part of the experiment, the pick-up Behringer dual-diaphragm condenser microphone B-2 PRO with switchable cardioid, omnidirectional, or figure-of-eight pickup patterns was used—see the directional patterns from the manufacturer's specification sheet in **Figure 6**.

Subsequently, the spectral properties of the recorded noise signals were analyzed using the mentioned three microphone pickup patterns. The obtained results are presented for visual comparison in **Figure 7** and summarized in numerical form in **Table 1**; the output statistical parameters of the supplementary spectral features are shown in **Figure 8**.

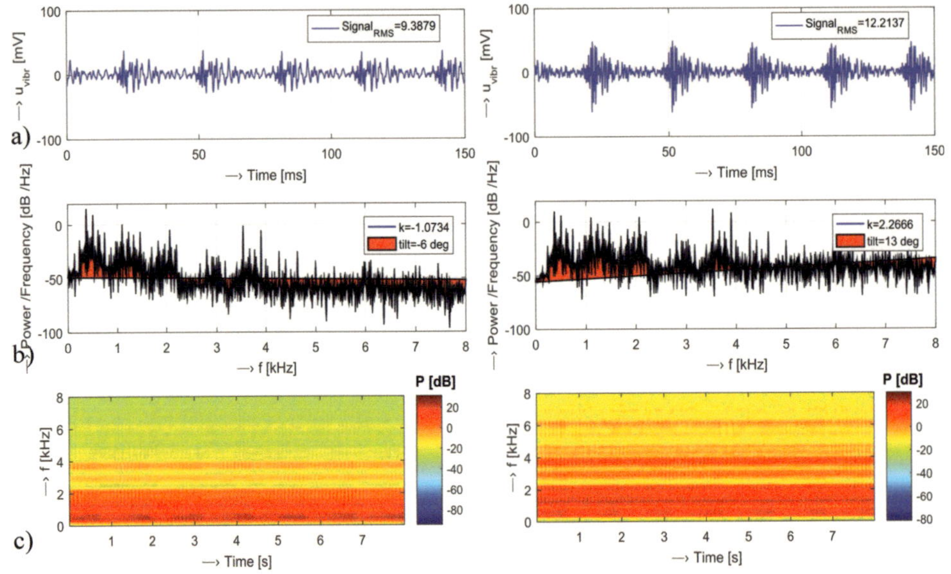

Figure 4.
The vibration signal picked up by the SB-1 sensor without/with the applied shelving filter (left/right set of graphs): selected 150-ms ROI of the signal together with the calculated RMS value (a), corresponding periodogram including the spectral decrease-tilt (b), and spectrogram calculated from the whole 8-s duration of the vibration signal (c); $Q = 0.115$, $f_c = 120$, $G = 30$, *and* $f_s = 16$ *kHz.*

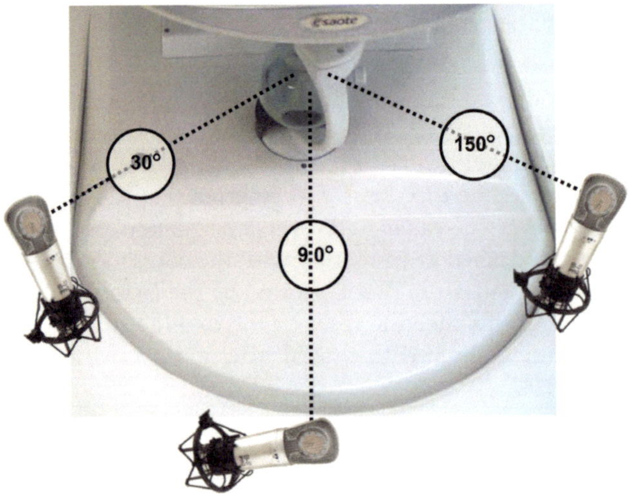

Figure 5.
Principal arrangement of acoustic noise recording in the vicinity of the scanning area of the open-air MRI device Opera: the pickup microphone situated at 30, 90, and 150°.

3.3 Mapping of the acoustic noise SPL in the MRI device vicinity

The acoustic noise SPL was measured using the multifunction environment meter Lafayette DT 8820. In the first step, the dependence of the SPL noise values on the distances D_X was mapped. The measuring device was located successively at the distances of {45, 50, 55, 60, 70, 80, 90} cm from the central point of the scanning area, at the height of 85 cm from the floor (between both gradient coils), and in the direction of 30° from the left corner near the temperature stabilizer, producing majority of the background noise SPL_0—see the experiment

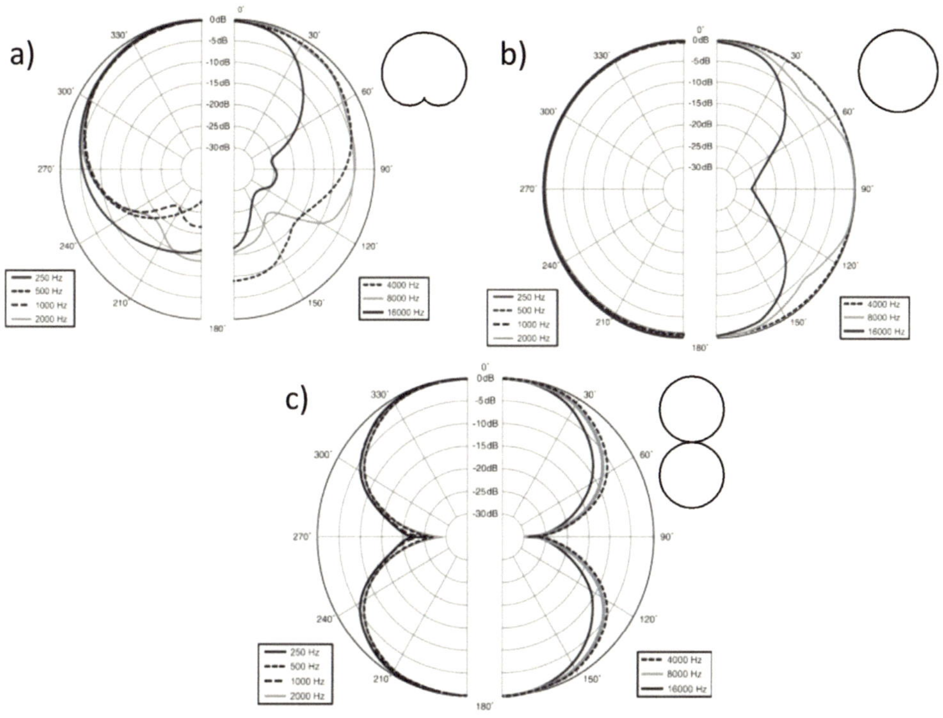

Figure 6.
Example of directional patterns: cardioid (a), omnidirectional (b), and figure-of-eight (c) for the Behringer condenser microphone B-2 PRO.

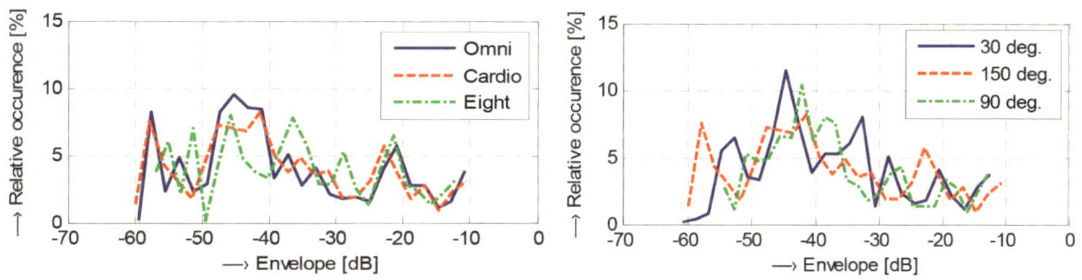

(a) (b)

Figure 7.
Comparison of spectral envelope values in [dB] of the noise signals with different directional patterns of the pickup microphone placed at different positions: histograms for omnidirectional, cardioid, and figure-of-eight patterns—signals recorded at 90° (a) and histograms for signals recorded at 30, 90, and 150°—with the cardioid directional pattern (b).

Microphone pickup pattern/ position	At 30°		At 90°		At 150°	
	Signal $_{RMS}$ [−]	S_{tilt} [deg]	Signal $_{RMS}$ [−]	S_{tilt} [deg]	Signal $_{RMS}$ [−]	S_{tilt} [deg]
Omnidirectional	15.3	−16	13.5	−15	14.2	−13
Cardioid	15.2	−11	13.3	−10	14.0	−4
Figure-of-eight	14.1	−18	13.1	−13	13.0	−9

Table 1.
Comparison of the noise spectral parameters of the recordings picked up by the microphone with different directional patterns placed at different positions.

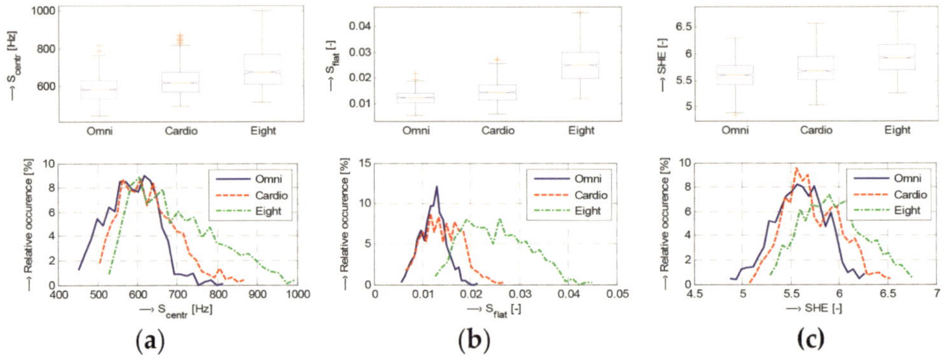

Figure 8.
Supplementary spectral properties of the recorded noise signals with different directional patterns of the pickup microphone—(a) omnidirectional, (b) cardioid, and (c) figure-of-eight; box-plots of the basic statistical parameters in the upper graphs, corresponding histograms of values of the spectral centroid, flatness, and Shannon entropy (in the lower set of graphs); signal recorded at 90°.

arrangement photo in **Figure 9**. Comparison of the resulting SPL values obtained during execution of two basic SE and GE types of the MR scan sequences with the background noise SPL (with no sequence running) is presented in the graphs of **Figure 10**.

3.4 Main measurement experiments with the open-air MRI device

Within the scope of our main experiments, the baseline measurement and recording of the vibration and noise signals were carried out during the execution of the MR scan sequences. For noninvasive testing of the subject/object, usually two basic classes of scan sequences are used to take MR images of human body parts with high quality:

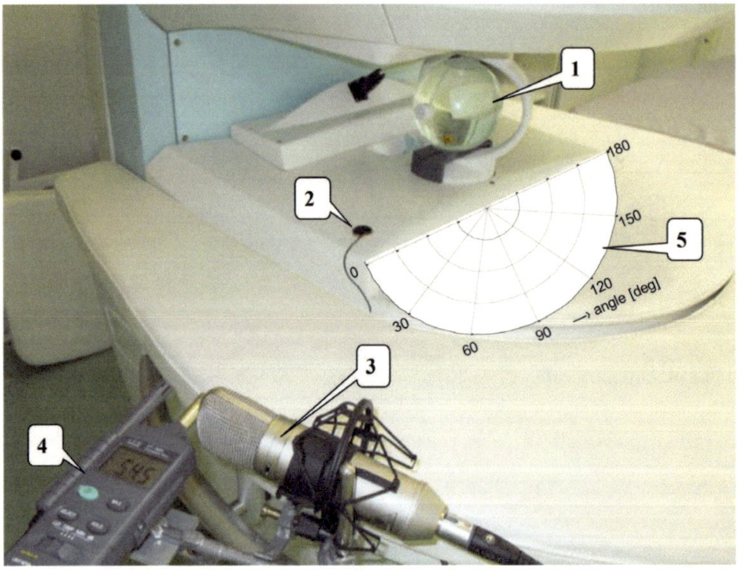

Figure 9.
Arrangement photo of SPL noise measurement and parallel recording of noise and vibration signals of the open-air MRI device Opera: (1) RF knee coil with a spherical water phantom, (2) vibration sensor, (3) pick-up microphone, (4) SPL noise meter, and (5) principal angle diagram of the scanning area.

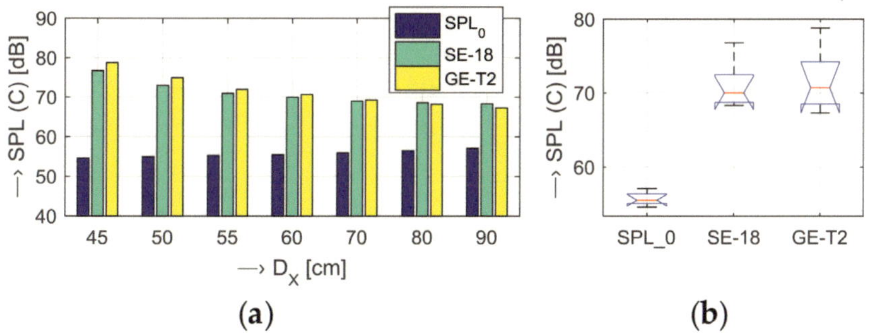

Figure 10.
Mapping of the acoustic noise SPL at different distances D_X = {45, 50, 55, 60, 70, 80, 90} cm from the middle of the scanning area of the MRI device for SE/GE sequences: (a) comparison of the SPL values with those of the background noise (SPL_0) and (b) box-plot of their basic statistical parameters.

- high-resolution (Hi-Res) sequences using the basic SE/GE MRI scan methods [16],
- special 3D sequences used for building or reconstruction of 3D models of biological or botanical issues [17].

Five types of MR scan sequences were tested in total in the investigated MRI device Opera: SE 18 HF, SE 26 HF, GE T2 (as a typical representative of the "Hi-Res" class), SS-3Dbalanced, and 3D-CE [15]. For each of these scan sequences, different settings of the scan parameters are analyzed:

- orientation of scan slices T_{ORIENT} = {Coronal, Sagittal, Transversal}—see visualization of the energy features of the vibration and noise signals in **Figure 11**,

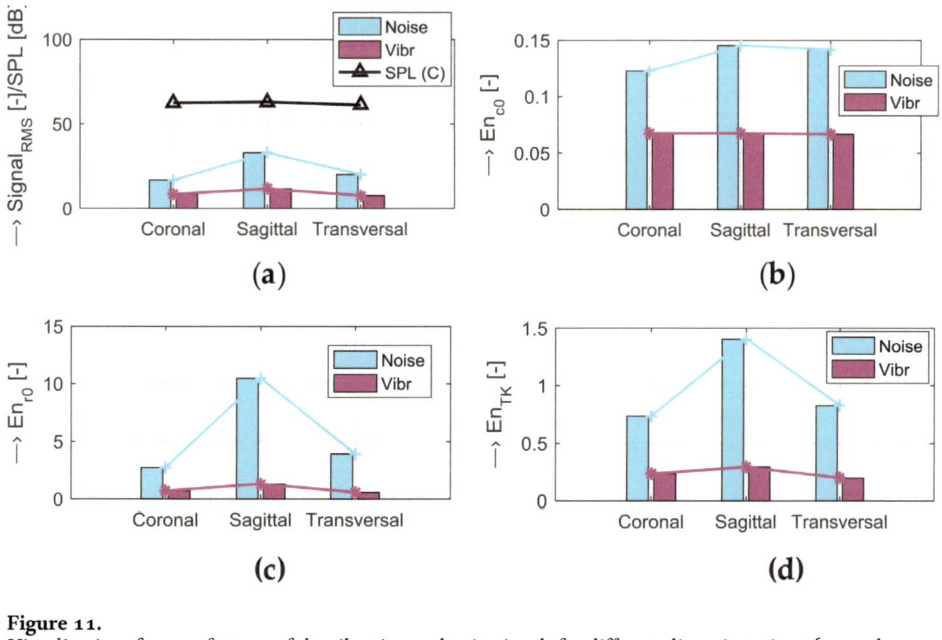

Figure 11.
*Visualization of energy features of the vibration and noise signals for different slice orientations: {coronal, sagittal, transversal}: (a) signal RMS together with noise SPL values, (**b**) mean En_{co}, (c) mean En_{ro}, and (d) mean En_{TK}; used Hi-Res SE scan sequence with TE = 18 ms and TR = 500 ms.*

- echo times T_{TE} = {18, 22, 26} ms—compare the numerical results in **Table 2**,
- repetition time T_{TR} = {60, 100, 200, 300, 400, 500} ms—documented by comparison of the basic statistical parameters calculated from the vibration and noise signals in **Figure 12**,
- mass of the object inserted in the MRI device scanning area {testing phantom/ lying person}—see graphical comparison of the mean values of the energy and basic spectral properties of the vibration signal in **Figure 14**.

The slice orientations as well as the TE and TR parameters were set manually to perform measurement and comparison in the range enabled by the current sequence [15]. Practical realization of the last part of the experiment consists in placing a testing phantom or a head and a neck of a lying person in the RF scan coil between the upper and lower gradient coils of the MRI device. While the total weight of the used testing phantom in the first part of the experiment was 0.75 kg, the weighs of one male and one female voluntary person lying on the patient bed of the MRI device were approx. 80 and 55 kg.

The multisignal measurement comprised real-time recording of the vibration signal by the piezoelectric sensor located inside the scanning area of the investigated

Sequence[1]	**Vibrations (SB-1)**				**Noise**[2] **SPL (C) [dB]**
	Signal $_{RMS}$[−]	***En***$_{TK}$ [−]	***En***$_{co}$[−]	***En***$_{ro}$[−]	
TE = 18 ms	31.5 (1.53)	4.32 (0.67)	0.044 (0.002)	23.04 (4.7)	61.5
TE = 22 ms	34.6 (2.11)	4.96 (1.02)	0.040 (0.003)	24.03 (8.5)	62.5
TE = 26 ms	36.0 (2.27)	5.75 (0.85)	0.055 (0.004)	24.40 (9.3)	63.0

[1]*Used Hi-Res SE-HF scan sequences with TR = 500 ms and sagittal orientation.*
[2]*Measured at the distance of* D_X *= 60 cm and the angle of 30°,* SPL_0 *= 56 dB.*

Table 2.
Comparison of the mean energetic parameters of the vibration signal and the acoustic noise SPL (together with std. values in parentheses) for different settings of the TE time.

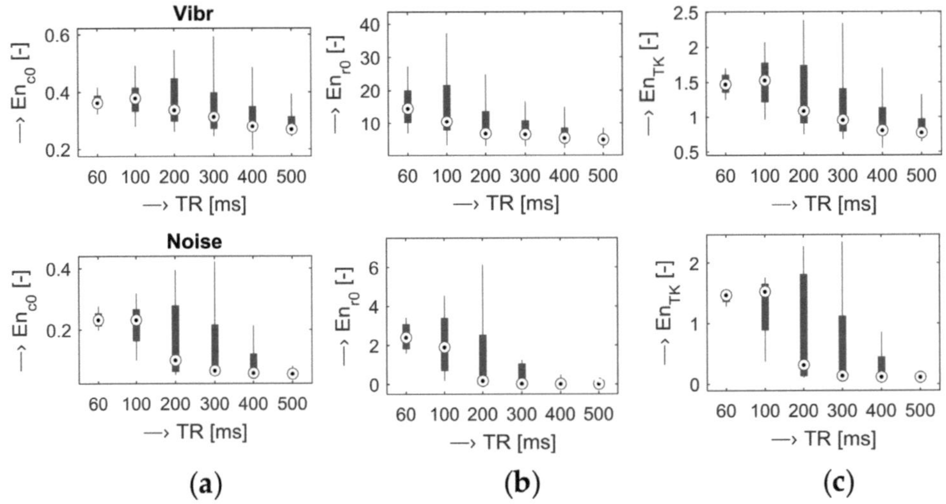

Figure 12.
Visualization of energetic relations of the vibration (upper set of graphs) and noise (lower set) signals for different TR times; {60, 100, 200, 300, 400, 500} ms—basic statistical parameters of: (a) En_{co}*, (b)* En_{ro}*, and (c)* En_{TK}*; used Hi-Res GE-T2 sequences with TE = 22 ms and sagittal orientation.*

MRI device and of the acoustic noise signal using the microphone in its proximity, and the additional measurement to check the noise SPL. In this part of the measurement, the microphone stand with the Behringer dual-diaphragm condenser microphone B-2 PRO was placed together with the SPL meter at the distance of D_X = 60 cm, and the 140-mm diameter spherical testing phantom filled with doped water [15] was placed inside the knee RF coil. The SB-1 sensor [12, 18] was used to pick up the vibration signal inside the scanning area of the MRI Opera device. Practical position of the sensing disc was on the surface of the plastic holder of the bottom gradient coils, as can be seen in the arrangement photo in **Figure 9**. The stored recordings were further processed in order to evaluate and compare the measured signal properties. All the noise and vibration signals were recorded with the help of the Behringer Podcast Studio equipment. The signals with duration of about 15 s sampled at 32 kHz were next processed in the sound editor program Sound Forge 9.0a.

3.5 Analysis of the influence of the scan parameters on the time duration and the quality factor of the MR images

The chosen type of the scanning sequence and the values of the resulting basic scan parameters (TR and TE) have significant influence on the scanning time. These parameters can also be changed manually, but their final values depend on the setting of the other scan parameters—number of slices, slice thickness, number of used accumulations N_{ACC} of the free induction decay (FID) signal [8, 16], etc. Practical demonstration of the acquired MR images with increasing quality factor (Q_F) shows greater range of visible details in the images for three different MR scans of the human vocal tract in **Figure 15**.

The console program "ESAMRI" of the MRI device control software [15] was used to carry out the following two parts of the analysis and comparison:

1. Influence of the basic setting of scan parameters on the final quality factor of MR images and on the time duration T_{DUR} of the scan sequence execution for:

   - different slice thickness of {2, 2.5, 3, 4, 4.5, 5, 10} mm—the predicted Q_F values are presented in **Table 3** for the scan sequence Hi-Res SE18 HE,

   - different repetition times of {60, 100, 200, 300, 400, 500} ms together with N_{ACC}—see visualization of the graphical results using the "Hi-Res" sequences of SE and GE types in **Figure 16**, and T_{DUR} values in **Table 4** for both Hi-Res sequences types,

   - increased number of applied accumulations of the FID signal: N_{ACC} = {1, 2, 3, 4, 5, 6, 7, 8, 10, 16}—the predicted values of Q_F and T_{DUR} are shown numerically in **Table 5** for the scan sequence Hi-Res SE18 HE.

Parameters[1]	Slice thickness [mm]						
	2	2.5	3	4	4.5	5	10
Q_F [−]	17	21	26	34	38	43	85

[1]*T_{DUR} = 1 min 39 sec in all cases.*

Table 3.
Influence of the slice thickness on the predicted quality factor of the MR image and on the time duration for the scan sequence Hi-Res SE18 HE (TR = 500 ms, N_{ACC} = 1).

N_{ACC} [−]	TR [ms]					
	60	100	200	300	400	500
1	0:14	0:22	0:41	1:09	1:20	1:39
8	1:35	2:37	5:12	7:46	10:20	12:55
16	3:08	5:11	10:20	15:29	20:38	25:47

Table 4.
Dependence of the time duration T_{DUR} [min:sec] on setting of TR and N_{ACC} parameters—merged values for both Hi-Res sequences of SE and GE types; slice thickness = 4.5 mm.

Parameters	N_{ACC} [−]									
	1	2	3	4	5	6	7	8	10	16
Q_F [−]	14	20	24	28	31	34	37	40	44	56
T_{DUR} [min:sec]	0:14	0:26	0:37	0:49	1:00	1:12	1:24	1:35	1:58	3:08

Table 5.
Influence of the number of FID signal accumulations on the predicted quality factor of the MR image and on the time duration for the scan sequence Hi-Res SE18 HE (TR = 60 ms and slice thickness = 10 mm).

Parameters	N_{ACC} [−]					
	1	2	3	4	8	16
Q_F [−]	59 (102)	84 (144)	103	118 (204)	167	237
T_{DUR} [min:sec]	3:14 (5:36)	6:25 (11:04)	9:37	12:48 (22:00)	25:34	51:05

Table 6.
Influence of the number of FID signal accumulations on the predicted quality factor of the MR image and on the time duration for the scan sequence SS-3D balanced (TE = 10 ms and TR = 20 ms) and 3D phases = 24 (for 42 phases, the values are in parentheses).

Parameters	N_{ACC} [−]					
	1	2	3	4	8	16
Q_F [−]	134 (79)	189 (122)	231	267 (137)	378	534
T_{DUR} [min:sec]	1:04 (9:53)	2:00 (19:44)	2:56	3:52 (29:35)	7:36	15:04

Table 7.
Influence of the number of FID signal accumulations on the predicted quality factor of the MR image and on the time duration for the scan sequence 3D-CE (TE = 30 ms and TR = 40 ms) and 3D phases = 8 (for 72 phases the values are in parentheses).

2. Comparison of the predicted Q_F and T_{DUR} values for "3D" types of MR scan sequences—numerical matching of the results for the changed number of FID signal accumulations and different number of 3D phases using:

   - the SS-3D-balanced 10 sequences—see the values in **Table 6**,
   - the 3D-CE 30 scan sequence (see **Table** 7).

4. Discussion of the obtained results

The performed calibration and frequency response linearization of the piezoelectric vibration sensor enables precise pick-up of vibration signals in the environment of a weak stationary magnetic field and a high-voltage RF signal disturbance that is observed in the scanning area of the MRI device.

Our measurements have shown an inverse relationship between the diameter of the used sensor and the minimum frequency of the vibration picked up from the measured surface. The sensor HM692 with a massive aluminum microphone capsule used in phonocardiography had the lowest sensitivity and caused the greatest decrease of the maximum frequency. The calibration of the SB2 sensor was carried out in parallel for both pickup elements. The measured frequency responses *SB2a,b* are practically identical with nonlinear decrease in the range of low frequencies from 35 to 100 Hz—see the frequency responses in **Figure 3a**. In 3D scanning of the human vocal tract [4, 5, 19], the MRI device generates the acoustic noise of frequencies in the range from 25 Hz to 3.5 kHz that is similar to the basic frequency range of speech signals. For this reason, the SB-1 sensor was chosen for its greatest size allowing the best low-frequency sensitivity.

Comparison of noise spectral properties recorded for different types of directional patterns of the pickup microphone yields the best recording conditions for the cardioid pattern (minimum spectral decrease as shown by the obtained results in **Table 1**). On the other hand, dispersion of the spectral envelope values is similar for all three analyzed pattern types as can be seen in histograms in **Figure 7a**. Comparison of different microphone positions has shown that at 30°, the background noise from the MRI temperature stabilizer degrades the recording (see the signal RMS values in **Table 1**) and the direction of 150° is a bit unnatural from the point of view of an examined person lying in the MRI scanning area. Therefore, the direction chosen as the best for noise and speech signal recording was in the main horizontal axis of the MRI device (at 90°). In addition, at this position, the lowest values of the noise signal RMS were measured and the smallest dispersion of the spectral envelopes was observed—see the green dash-dot line in **Figure 7b**.

The results of a detailed measurement of the acoustic noise intensity at different distances from the central point of the scanning area for the SE and GE "Hi-Res" sequences are presented in **Figure 10**. The GE sequence produces noise with a slightly higher intensity, then the SE one (approx. 3-dB difference in the nearest location of 45 cm from the center of the scanning area) and variation of the SPL values depending on the measuring distance is also greater as seen in the box-plot graph in **Figure 10b**. The minimum distance was set to 45 cm in order to eliminate interaction of metal parts of the SPL meter with the static magnetic field of the MRI device. If the SPL meter was placed near the center, the field homogeneity would be disrupted and the warning message on the MRI control console would be followed by disabling to run any scan sequence by the software system [14]. The maximum measuring distance was set to 90 cm where the measured MRI noise was masked by the background noise originating from the temperature stabilizer. In the middle of the investigated measuring distances, the SPL values were similar for both types of MR scan sequences, so the working distance of 60 cm was used for all further measurements.

Next investigation of the recorded vibration and noise signals was aimed at the influence of the choice of the slice orientation on the energy of the produced vibration and noise signals. This effect is large—the maximum can be found in the sagittal plane and the minimum in the transversal plane for the vibration signals, and in the coronal plane for the noise signals—see the column charts in **Figure 11**. Therefore, the remaining experiments used only the sagittal orientation.

In accordance with our previous research [12, 18] the current experiments confirm the influence of the TE and TR times on the vibration and acoustic noise properties. The TE time extension causes fall of the final signal energy as documented by raised all the four determined vibration energetic parameters as well as the achieved SPL noise values in **Table 2**. The influence of the TR time determining the basic dominant frequency can be seen in box-plot graphs in **Figure 12**. This visualization of the basic statistical parameters obtained from analysis of vibration and noise signals shows the highest values of all energetic parameters for the shortest TR times (60 or 100 ms).

Comparison of energetic relations of the vibration and noise signals for different sequence types brings ambiguous results and shows only small differences—see three bar-graphs in **Figure 13**. The 3D sequence "SS-3Dbalanced" differs from the remaining sequence types by reverse behavior: while the En_{c0} and En_{r0} parameters indicate the minimum values, the En_{TK} achieves the maximum ones (see the graph in **Figure 13c**). This situation can be caused by the minimum settings of the TE and TR times that were used for the "Hi-Res" types to be comparable with the "3D" types with slightly atypical values being out of the normal range of use although the control software enables their setting [15].

Next comparison of energetic relations of the vibration and noise signals for different objects placed in the scanning area of the MRI device shows a relatively high effect of the mass put upon the bottom plastic holder of the gradient coils. The effective weight of the person exerting a pressure on the bottom plastic holder of the gradient coils attenuates the vibration pulses partially. The mass effect is demonstrated by increase of the vibration signal energy based on En_{c0} parameter with its maximum for the lying male person with the weight of 80 kg (see the bar-graph in **Figure 14a**). It is also demonstrated in the spectral properties of the vibration

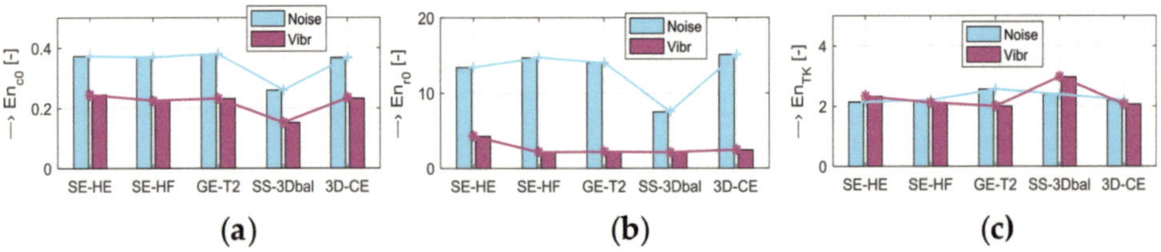

Figure 13.
Comparison of energetic relations of vibration and noise signals for different sequence types: {Hi-Res SE-HE, Hi-Res SE-HF, Hi-Res GE-T2, SS-3Dbal, 3D-CE}: (a) mean En_{co}*, (b) mean* En_{ro}*, and (c) mean* En_{TK}*; in all cases, the sagittal slice orientation was used.*

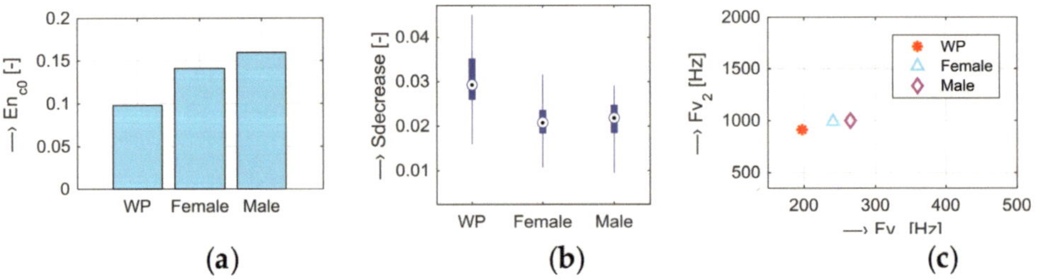

Figure 14.
Comparison of mean values of the energy and basic spectral properties of the vibration signal for different objects placed in the scanning area of the MRI device: (a) energy En_{co}*, (b) box-plot of basic statistical properties for the spectral decrease values, and (c) mutual values of the frequencies* F_{V_1} *and* F_{V_2}*; used Hi-Res SE-HF scan sequences with TE = 18 ms, TR = 400 ms, and sagittal orientation.*

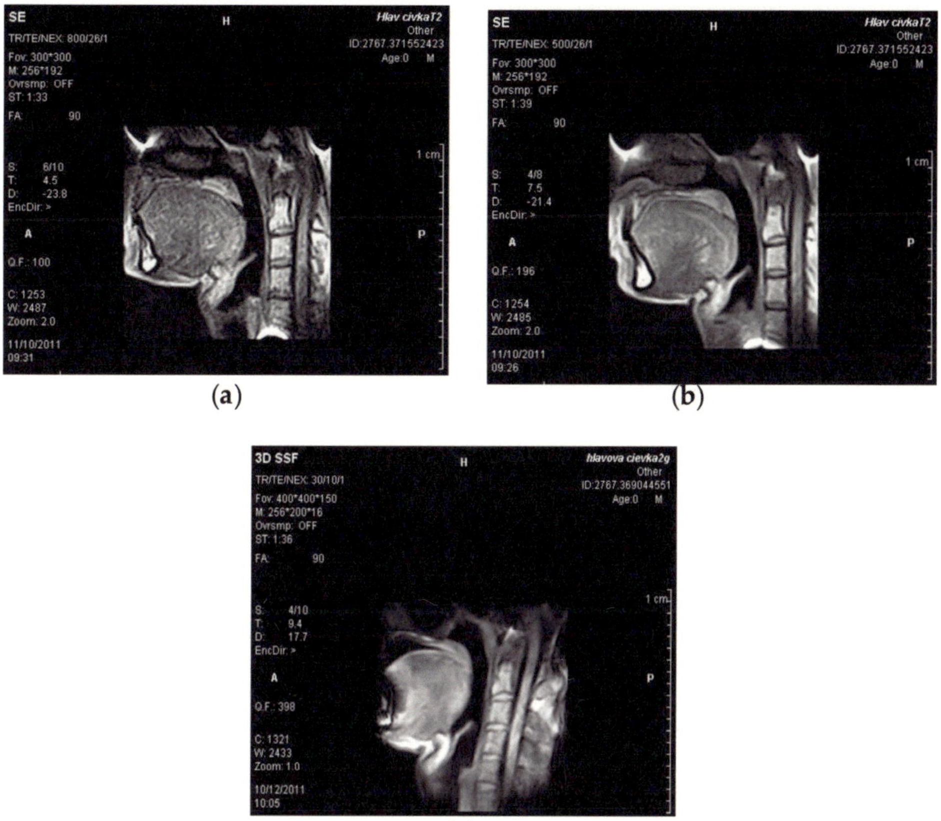

Figure 15.
Examples of MR images of the human vocal tract obtained with different values of the quality factor: (a) scan sequence Hi-Res SE26 HF (TR = 500), slice thickness = 4.5 mm, Q_F = 100, (b) scan sequence Hi-Res SE26 HF (TR = 500), slice thickness = 7.5 mm, Q_F = 196, and (c) scan sequence 3D SSF 30 (TR = 10), slice thickness = 9.4 mm, and Q_F = 398.

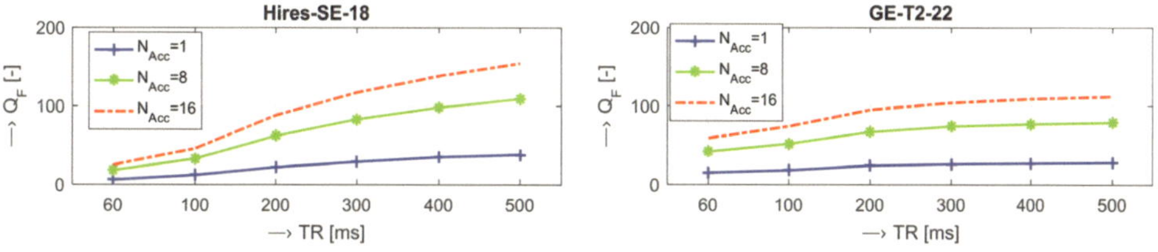

Figure 16.
Influence of the TR time and the number of FID signal accumulations on the predicted image quality factor for the Hi-Res sequences of—(a) SE and (b) GE type; slice thickness = 4.5 mm.

signal as shown by lower spectral decrease in **Figure 14a** and by shift of the first two dominant frequencies toward higher values—see the mutual $F_{V1,2}$ position in **Figure 14c**.

Results of the preliminary analysis of influence of the slice thickness document that its increase has a positive effect on the predicted quality factor of MR images—compare the values in **Table 3**. Next comparison shows a positive influence of increase in the TR time on the quality factor, and this effect is more pronounced when using the SE sequence type—see the left graph in **Figure 16**. This figure also documents significant dependence between the applied number of FID signal accumulations and the predicted Q_F value. Also, in this case, the increase of Q_F is more distinctive for the SE sequences. Values in **Table 4** describe the influence of TR and N_{ACC} values on the final time duration of the executed scan sequence. While the increased TR causes only moderately greater overall time duration, the changed N_{ACC} parameter has comparably higher influence on the final time duration. This effect is also shown in a detailed comparison of numerical results for different N_{ACC} values in **Table 5**. For the "Hi-Res" sequence types, the increase of the parameter N_{ACC} from 2 to 16 results in about 2.8 times greater value of Q_F but 6 times greater than that of T_{DUR}. For the "3D" sequence types, the increase of the resulting time duration is also affected by the choice of the number of 3D phases (equivalent to the number of slices with selection of the slice thickness for the "Hi-Res" sequences) in parallel as shown in **Tables 6 and 7**.

5. Conclusions

Acoustic noise measurement in the vicinity of the investigated open-air MRI device yielded the maximum sound pressure level of about 82 dB(C) at the distance of 45 cm from the central point of the MRI scanning area for the GE scan sequence with short TE and TR times and the sagittal orientation of scan slices. For examination of other parts of the human body (leg, arm, etc.), the head is not inserted directly between the upper and the lower gradient coils, so the noise level is much lower as documented for different distances in **Figure 10**. Finally, the scanning times for the mostly used 3D or Hi-Res sequences are in general less than 15 minutes (typically about 3–5 minutes depending on the chosen number and thickness of the slices)—exposition of the examined person and his/her hearing system to the noise and vibration is not significant.

If there is need for more detailed MR images with higher quality factor Q_F (e.g., in scanning of particular parts of the human brain, the eye, the middle and inner ear, etc.), the time duration T_{DUR} can be much longer (more than half an hour). In such a case, the long exposition to the vibration and acoustic noise may impose great physiological and psychological stress on the patient. Therefore, these scan parameters should be chosen only in the urgent cases.

The results of the performed measurements are useful for precise description of the process of the mechanical vibration excitation and the acoustic noise radiation in the scanning area and in the vicinity of the MRI device. The measurement results and comparisons with a similar low-field MRI tomograph can be used in optimization of the acoustic noise suppression in the speech recorded parallel with application of MRI scanning for 3D modeling of the human vocal tract [19].

Acknowledgements

This work was funded by the Slovak Scientific Grant Agency project VEGA 2/0001/17 and the Ministry of Education, Science, Research, and Sports of the Slovak Republic VEGA 1/0905/17, and the Slovak Research and Development Agency, project no. APVV-15-0029.

Conflict of interest

The authors declare no conflict of interest.

Author details

Jiří Přibil[1]*, Anna Přibilová[2] and Ivan Frollo[1]

1 Institute of Measurement Science, SAS, Bratislava, Slovak Republic

2 Institute of Electronics and Photonics, Faculty of Electrical Engineering and Information Technology, SUT, Bratislava, Slovak Republic

*Address all correspondence to: umerprib@savba.sk

References

[1] He Z, He W, Wu J, Xu Z. The novel design of a single-sided MRI probe for assessing burn depth. Sensors. 2017;**17**: 526. DOI: 10.3390/s17030526

[2] Panych LP, Madore B. The physics of MRI safety. Journal of Magnetic Resonance Imaging. 2018;**47**:28-43. DOI: 10.1002/jmri.25761

[3] Moelker A, Wielopolski PA, Pattynama PMT. Relationship between magnetic field strength and magnetic-resonance-related acoustic noise levels. Magnetic resonance materials in physics. Biology and Medicine. 2003;**16**: 52-55. DOI: 10.1007/s10334-003-0005-9

[4] Mainka A, Platzek I, Mattheus W, Fleischer M, Müller AS. Three-dimensional vocal tract morphology based on multiple magnetic resonance images is highly reproducible during sustained phonation. Journal of Voice. 2017;**31**:504. e11-5504.e20. DOI: 10.1016/j.jvoice.2016.11.009

[5] Kuortti J, Malinen J, Ojalammi A. Post-processing speech recordings during MRI. Biomedical Signal Processing and Control. 2018;**39**:11-22. DOI: 10.1016/j.bspc.2017.07.017

[6] Seidman MD, Standring RT. Noise and quality of life. International Journal of Environmental Research and Public Health. 2010;7:3730-3738

[7] Diedrichsen J, Balsters JH, Flavell J, Cussans E, Ramnani N. A probabilistic MR atlas of the human cerebellum. NeuroImage. 2009;**46**:39-46. DOI: 10.1016/j.neuroimage.2009.01.045

[8] Liang ZP, Lauterbur PC. Principles of Magnetic Resonance Imaging: A Signal Processing Perspective. New York: Wiley-IEEE Press; 2000. 416 p. ISBN: 978-0-780-34723-6

[9] Rimell AN, Mansfield NJ, Paddan GS. Design of digital filters for frequency weightings (A and C) required for risk assessments of workers exposed to noise. Industrial Health. 2015;**53**:21-27. DOI: 10.2486/indhealth.2013-0003

[10] Fraden J. Handbook of Modern Sensors: Physics, Designs, and Applications. 4th ed. New York: Springer; 2016. 663 p. ISBN: 978-1-4939-0040-4

[11] Mechefske CK. Vibration in MRI scanners. In: Al-Jumaily A, Alizad A, editors. Biomedical Applications of Vibration and Acoustics in Therapy, Bioeffects and Modeling. New York: ASME Press; 2008. pp. 329-349. ISBN: 978-0-7918-0275-5

[12] Přibil J, Přibilová A, Frollo I. Comparison of mechanical vibration and acoustic noise in the open-air MRI. Applied Acoustics. 2016;**105**:13-23. DOI: 10.1016/j.apacoust.2015.11.013

[13] Zölzer U. Digital Audio Signal Processing. 2nd ed. Chichester: John Wiley & Sons; 2008. ISBN: 978-0-470-99785-7

[14] Boudraa AO, Salzenstein F. Teager–Kaiser energy methods for signal and image analysis: A review. Digital Signal Processing. 2018;**78**:338-375. DOI: 10.1016/j.dsp.2018.03.010

[15] SpA E. E-Scan Opera. Genoa: User's Manual. Revision A; 2008

[16] Bernstein MA, King KF, Zhou XJ. Handbook of MRI Pulse Sequences. Burlington: Elsevier Academic Press; 2004. 1040 p. ISBN: 978-0-12-092861-3

[17] Wellard RM, Ravasio JP, Guesne S, Bell C, Oloyede A, Tevelen G, et al. Simultaneous magnetic resonance imaging and consolidation measurement

of articular cartilage. Sensors. 2014;**14**: 7940-7958. DOI: 10.3390/s140507940

[18] Přibil J, Přibilová A, Frollo I. Mapping and spectral analysis of acoustic vibration in the scanning area of the weak field magnetic resonance imager. Journal of Vibration Acoustic Transaction ASME. 2014;**136**:051005–01-051005–051010. DOI: 10.1115/1.4027791

[19] Přibil J, Přibilová A, Frollo I. Analysis of acoustic noise and its suppression in speech recorded during scanning in the open-air MRI. In: Ahmed N, editor. Advances in Noise Analysis, Mitigation and Control. InTech. Croatia: Rijeka; 2016. pp. 205-228

Chapter 7

Vibration Analysis and Control in the Rail Car System Using PID Controls

Ilesanmi Afolabi Daniyan and Khumbulani Mpofu

Abstract

The PID classic control systems are often employed for rail car systems to reduce the vibrations and disturbance rate during movement. In this study, the dynamic modeling and simulation of PID controls for rail car systems were carried out. Using 9 degrees of freedom, the modeling process comprises the representation of the rail car system and the rail track followed by the generation of equations of motion as well as differential equations for the rail car body, wheel sets and bogie. The represented systems are simulated in the MATLAB Simulink 2018 environment based on the equations of motion generated, and subsequently vibration analysis was carried out. The PID control system tuned according to the Nichols-Ziegler rule was introduced to minimize the vibrations and disturbance rate. The performance of the control and the rail car system in terms of the input step response, bandwidth, frequency, phase margin, frequency and input and output rejections was evaluated. The control system demonstrated significant robustness in providing the required active control for the system, while there was improved stability and reduction in noise and vibration under control action of the PID, thus improving ride comfort.

Keywords: advanced controls, modeling, rail car, simulation, vibration

1. Introduction

The rail car system uses steel wheels moving on steel rails, and its suspension system comprises the bogie, primary and secondary suspension elements as well as springs and dampers. The rail car has the power bogies in the front and rear positions each having four wheels arranged in pairs and rigidly connected through an axle. The sets of wheel, which rotates at the same speed, are connected to the bogie through the primary suspension system, which is harder and stiffer in order to minimize load disturbances and uneven weight distributions. This is to maintain a good balance of the rail car while moving along its track. The stiff primary suspension system connects the wheel set to the bogie, while the soft secondary suspension system connects the bogie to the rail car body in order to isolate it from disturbances that stem from rail irregularities, uneven track profiles and its associated vibrations. While the primary suspension system is designed to provide guidance and rail car stability amidst load and weight variations, the secondary suspension system is to enhance comfortable ride by isolating rail car body from rail track irregularities. These design requirements of both suspension systems are to dampen the effect of vibration and increase the overall system's performance [1, 2]. For a rail car, when vibrations are not kept within the permissible limits, the comfort, safety and health of passengers are at risk, and the time interval for maintenance activities for the components will likely increase. Hence, undesirable vibration in the rail car system is the cause of noise, considerable energy loss, reduction in the system's performance, fatigue or fracture of some component, instability of a moving rail car and displacement of the rail track, amongst others. Over the years, there are three main types of suspension systems often employed to check vibration in the rail car system, namely, the passive, semi-active and fully active suspension systems [3, 4]. The design requirements of the suspension system of the rail car are to provide support against the dynamic loads and weight of the railcar; prevent load and rail disturbances; isolate the rail car body disturbances that could offset its balance; prevent irregular motions such as bouncing, yawing, etc.; provide guidance along the rail track; and optimize the curving, braking and other car maneuvering performances. The passive suspension system, which consists of springs, absorbs and stores the energy absorbed, while the dampers mounted on each wheel act as the shock absorber that dissipates the energy stored in the spring and reduces the vibrations from the rail transmitted to the vehicle ([5–7]. The passive suspension system is cost-effective and driven by simple technology, but its demerit lies in the fact that it cannot measure some critical parameters such as the velocity, displacement, acceleration, etc. of the rail car system in real time in order to effect the needed adjustments to stabilize the rail car system; hence, it is rigid and cannot reach the compromise between alternate hard primary and soft secondary suspension systems amidst load and rail irregularities. On the other hand, the active suspension system uses sensors, comparators and actuator for measuring and monitoring some critical parameters that influence rail car stability in real time [8, 9]. The measured system's parameters are compared with the threshold already pre-set on the controller, which can guarantee comfortable ride. The steady-state errors generated are eliminated through adjustment and compensation for such errors, and as such, there is high-level damping without compromising the system's performance. The third category of suspension system is the semi-active suspension system, which employs a spring and controllable damper. The spring element stores the energy, while the controllable damper dissipates the energy stored. The semi-active suspension systems combine the features of the passive and active suspension systems such as the use of passive damper and an actively controlled spring. The merit of the semi-active suspension is that it is cost and energy effective [9–11].

Over the years, researchers have employed several approaches ranging from classic to advance systems to control the suspension system in order to minimize the effect of vibrations. Such approaches include the use of proportional-integral-derivative (PID) control, Fuzzy PID, Linear Quadratic Regulator vibration controller, Adaptive Neuro-Fuzzy Inference System control and magnetorheological dampers, amongst others [12–16].

2. Materials and method

The analysis of vibration and control in the rail car system starts with the schematic representation of the rail car and track system including their degrees of freedoms and subsequent generation of equations of motion. The modeling is done with the masses of the system (rail car, body and wheel sets) having 6 degrees of freedom in the longitudinal, lateral and vertical directions as well as in the roll, pitch

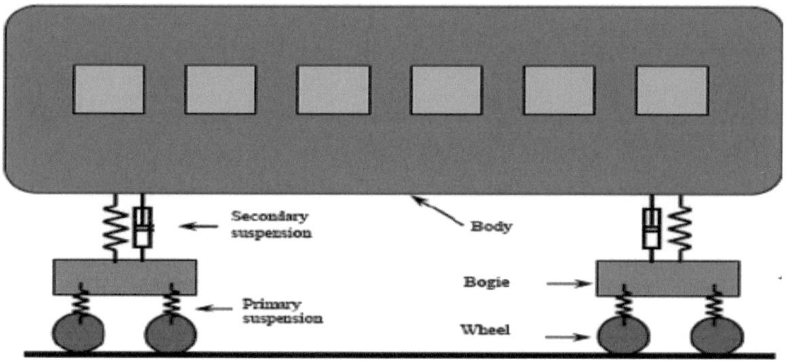

Figure 1.
The rail car and its suspension system [17].

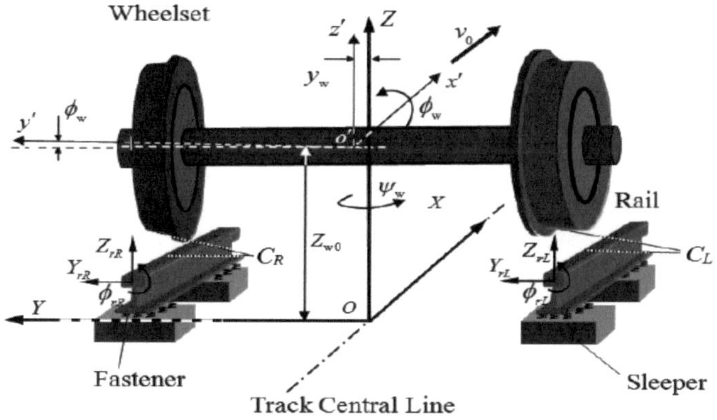

Figure 2.
The free body diagram for the rail car model [18].

and yaw rotational motions. **Figures 1** and **2** illustrate the free body diagram for the rail car and track system in their degrees of freedom.

In order to give a complete vibration analysis and control, the model design and simulation of the suspension systems are considered as both linear and non-linear systems. There are three forces acting on the rail car suspension system, namely, the spring force, rolling resistance and forces due to the wheel-track interactions. The spring force and the rolling resistance act on the rail car body with mass m_1 in the horizontal directions, while the spring force, rolling resistance and the forces due to the wheel-track interactions act on the bogies with mass m_2 due to load in the horizontal directions. The masses of the rail car and its bogies are represented by m_1 and m_2, respectively, while the bodies are connected to the rail car through the secondary suspension with couplings having stiffness K. If force F_d is the force generated as a result of wheel-track interaction, F_u is the control force and μ is the coefficient of rolling friction, then Newton's second law of motion holds; thus,

$$F = ma \tag{1}$$

$$\sum F_1 = K(x_1 - x_2) - \mu m_1 g\dot{x}_1 = m_1\ddot{x}_1 \tag{2}$$

$$\sum F_2 = F_d - K(x_1 - x_2) - \mu m_2 g\dot{x}_2 = m_2\ddot{x}_2 \tag{3}$$

Eq. (3) expresses the summation of the forces

$$\sum(F_1 + F_2) = m_1\ddot{x}_1 + m_2\ddot{x}_2 \tag{4}$$

$$\sum(F_1 + F_2) = M\ddot{x} \tag{5}$$

$$\sum(F_1 + F_2) = \sum(F_d + F_u) \tag{6}$$

The spring deflection is expressed as Eq. (7):

$$\delta = K(x_1 - x_2) \tag{7}$$

$$\sum(F_d + F_u) = M\ddot{x} + b\dot{x} + Kx \tag{8}$$

$$[F_d] + [F_u] = [M]\ddot{x} + [b]\dot{x} + [K]x \tag{9}$$

$$\text{Hence, } x = \left[x.....x^1\right]^T \tag{10}$$

$$\dot{x} = \frac{dx}{dt} = f(x(t), u(t), t) \tag{11}$$

where $x(t)$ is the state vector and a set of variables representing the configurations of the system. The modeling of the rail car system and its suspension system as well as the track system is done on the MATLAB Simulink 2018 environment (**Figure 3**).

The transfer function was used for representing the linear systems, and the inputs are the load changes, applied forces as well as the uneven track profiles, while the output of the system is the acceleration and displacement of the rail car body as well as the deflection of the suspension systems.

According to Sezer and Atalay [19], the vectors for the displacement, rail car disturbance and control forces are expressed as Eqs. (12)–(14), respectively:

$$\begin{aligned} x = \big[&X_r, L_r, V_r, \theta_r, \varnothing_r, \psi_r, X_{b1}, L_{b1}, V_{b1}, \theta_{b1}\varnothing_{b1}, \psi_{b1}, X_{b2}, L_{b2}, V_{b2}, \theta_{b2}\varnothing_{b2}, \psi_{b2},\\ &X_{b3}, L_{b3}, V_{b3}, \theta_{b3}\varnothing_{b3}, \psi_{b3}, X_{w1}, L_{w1}, V_{w1}, \theta_{w1}, \psi_{w1}, X_{w2}, L_{w2}, V_{w2}, \theta_{w2},\\ &\psi_{w2}, X_{w3}, L_{w3}, V_{w3}, \theta_{w3}, \psi_{w3}, X_{w4}, L_{w4}, V_{w4}, \theta_{w4}, \psi_{w4}, X_{w5}, L_{w5}, V_{w5},\\ &\theta_{w5}, \psi_{w5}, X_{w6}, L_{w6}, V_{w6}, \theta_{w6}, \psi_{w6}\big]T \end{aligned} \tag{12}$$

$$\begin{aligned} [F_d] = [&000000000000000000000000(k_{sL}Y_{r1})(k_{sV}(V_{r1} + V_{r2}))\\ &(k_{sL}R_1L_{r1} + k_{sV}(V_{r2} - V_{r1}))00(k_{sL}L_{r2})(k_{sV}(V_{r3} + V_{r4}))\\ &(k_{sL}R_1L_{r2} + k_{sV}a(V_{r4} - V_{r3}))00(k_{sL}L_{r3})(k_{sV}(V_{r5} + V_{r6}))\\ &(k_{sL}R_1L_{r3} + k_{sV}a(V_{r6} - V_{r5}))00\ (k_{sL}L_{r4})(k_{sV}(V_{r7} + V_{r8}))\\ &(k_{sL}R_1L_{r4} + k_{sV}a(V_{r8} - V_{r7}))00\ (k_{sL}L_{r5})(k_{sV}(V_{r9} + V_{r10}))\\ &(k_{sL}R_1L_{r5} + k_{sV}a(V_{r10} - V_{r9}))00(k_{sL}L_{r6})(k_{sV}(V_{r11} + V_{r12}))\\ &(k_{sL}R_1L_{r6} + k_{sV}a(V_{r12} - V_{r11}))0]^T \end{aligned} \tag{13}$$

$$\begin{aligned} [F_u] = [&0(U_{L1} + U_{L2} + U_{L3})(U_{V1} + U_{V2} + U_{V3} + U_{V4} + U_{V5} + U_{V6})\\ &(d_3(U_{L1} + U_{L2} + U_{L3})(p_2((U_{V2} + U_{V4} + U_{V6}) - (U_{V1} + U_{V3} + U_{V5})))\\ &L_o((U_{V5} + U_{V6}) - (U_{V1} + U_{V2})))(h_2U_{L1} - U_{L3}))0\\ &-(U_{L1})((U_{V1} + U_{V2}))(d_2U_{L1} - p_2(U_{V2} - U_{V1}))000(-U_{L2})\\ &(-U_{V3} + U_{V4}))(d_2U_{L2} - p_2(U_{V4} - U_{V3}))000(-U_{L3})(-U_{V5} + U_{V6}))\\ &(d_2U_{L3} - p_2(U_{V6} - U_{V5}))00000000000000000000000000000000]^T \end{aligned} \tag{14}$$

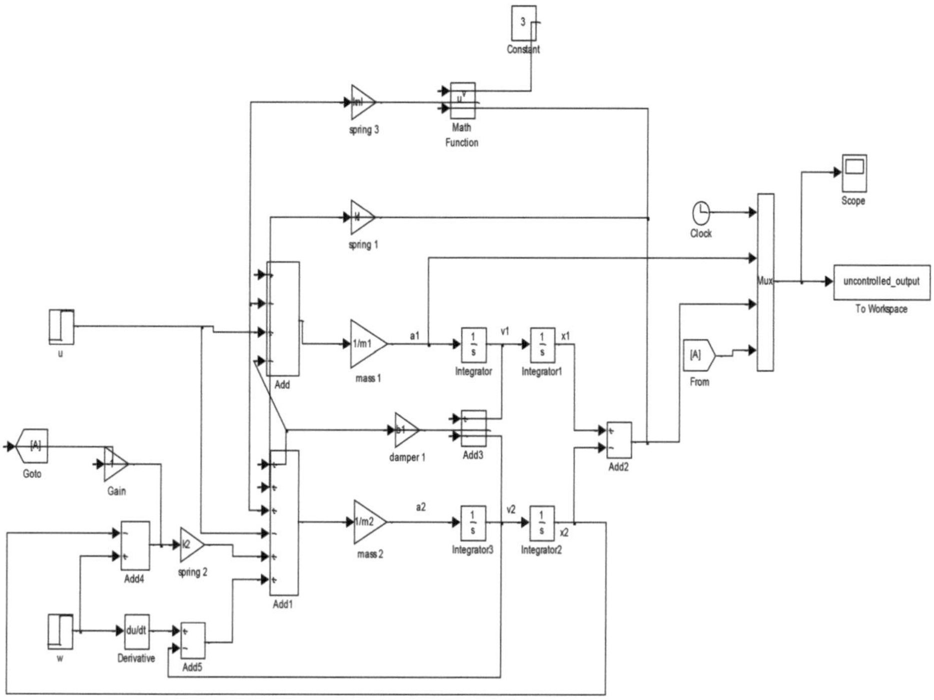

Figure 3.
Modeling of the rail car and its suspension system.

where.

$L_{r1}, L_{r2}, L_{r3}, L_{r4}, L_{r5}, L_{r6}$ are the lateral rail disturbance functions; $V_{r1}, V_{r3}, V_{r5}, V_{r7}, V_{r9}, V_{r11}$ are the vertical right rail disturbance functions; $V_{r2}, V_{r4}, V_{r6}, V_{r8}, V_{r10}, V_{r12}$ are the vertical left rail disturbance functions; X_r, L_r, V_r are the longitudinal, lateral and vertical displacements of the rail car body; X_{bi}, L_{bi}, V_{bi} are the longitudinal, lateral and vertical displacements of the rail car bogie; X_{wj}, L_{wj}, V_{wj} are the longitudinal, lateral and vertical displacements of the rail car wheel sets; $\theta_r, \varnothing_r, \psi_r,$ are the roll, pitch and yaw displacements of the rail car body; $\theta_{bi}, \varnothing_{bi}, \psi_{bi},$ are the roll, pitch and yaw displacements of the rail car bogie; $\theta_{wj}, \varnothing_{wj}, \psi_{wj},$ are the roll, pitch and yaw displacements of the rail car wheel set $(i = 1.....3; and\, j = 1.....6); p_1$ is the lateral distance between the vertical primary suspensions; p_2 is the lateral distance between the vertical secondary suspensions; d_1 is the vertical distance between the wheel set and the bogie mass centre; d_2 is the vertical distance between the bogie mass centre and lateral secondary suspension; d_3 is the vertical distance between the lateral secondary suspension and railcar body mass centre; L_o is the longitudinal distance between the bogies; and a is the lateral distance between the contact points of the wheel-rail.

The mathematical model can be obtained in either the state space or using the transfer function for necessary control actions. The control action is to improve the speed of response, system's balance and stability and to reduce the steady-state error as well as the amplitude of oscillations. Eqs. (15) and (16) express the form of the second-order systems:

$$\dot{y}(t) + \xi\omega_n\dot{y}(t) + \omega_n^2 y(t) = k_d\omega_n^2 u(t) \tag{15}$$

$$G(s) = \frac{X(s)}{F(s)} = \frac{1}{ms^2 + bs + k} = \frac{k_d\omega_n^2}{s^2 + 2\xi\omega_n s + \omega_n^2} = \frac{k_d\omega_n^2}{(s+\sigma)^2 + \omega_d^2} \tag{16}$$

where.

k_d is the dc gain which is the steady-state step response to the magnitude of the step input expressed as Eq. (17)

$$k_d = \frac{1}{k} \tag{17}$$

ω_d is the damped natural frequency expressed as Eq. (18)

$$\omega_d = \omega_n \sqrt{1 - \xi^2} \tag{18}$$

σ is the real part of the pole expressed as Eq. (19)

$$\sigma = \xi\omega_n \tag{19}$$

ω_n is the undamped natural frequency at which the system oscillates expressed as Eq. (20)

$$\omega_n = \sqrt{\frac{k}{m}} \tag{20}$$

ξ is the damping ratio which defines the rate or nature of amplitude of oscillation (Eq. (21))

$$\xi = \frac{b}{2\sqrt{km}} \tag{21}$$

The performance of the control system is measured by the settling time, delay time rise time, percent overshoot and the steady-state error.

The settling time t_s is the time it takes the system to fall within a certain percent (mostly 2%) of the steady-state value for a step input response expressed as Eq. (22):

$$t_s = \frac{-lnT_f}{\xi\omega_n} = 4\zeta = \frac{4}{\sigma} \tag{22}$$

On the other hand, the rise time is the time it takes the signal to change from a low value to a high value (say 10–90% or 0–100%). The time it takes the peak value to occur known as the time is expressed as Eq. (23):

$$t_p = \frac{\pi}{\omega_d} \tag{23}$$

The steady-state error $E(s)$ is the difference between the input reference signal $R(s)$ and the output signal $Y(s)$ expressed as Eq. (24):

$$E(s) = R(s) - Y(s) \tag{24}$$

Similarly, the delay time is the time required for the response to reach half the final value for the first time, while the percent overshoot is the percent by which the step response of the system exceeds the final steady-state value. It is a parameter that defines the instability of a system (Eq. (25)):

$$P_o = e^{-1} \frac{\zeta\pi}{\sqrt{1-\xi^2}} \tag{25}$$

Eq. (26) relates the percent overshoot to the damping ratio:

$$\xi = \frac{-lnP_o}{\sqrt{\pi^2 + ln^2(P_o)}} \tag{26}$$

The input parameters of the system are presented in **Table 1**.

2.1 The proportional-integral-derivative (PID) control

The PID control represents the proportional, integral and derivative controls. It is a form of classic control designed to automatically reduce rise and settling times, steady-state errors and percent overshoot. The block diagram of the PID controller is illustrated in **Figure 4**. The threshold value otherwise referred to as the reference point is pre-set on the controller while real-time measurement using sensors is taken. The error generated, which is the difference between the threshold value and the actual measurement, represents the deviation from the ideal process; thereafter, the actuator effects real-time control to adjust process variables. The output signal of the PID controllers often responds to changes over time with respect to the actual

S/N	Parameter	Notation	Value	Unit
1	Average mass of the rail car	M_1	50,500	kg
2	Average mass of bogie	M_2	2410	kg
3	Mass of primary suspension system	M_p	30,000	kg
4	Mass of secondary suspension system	M_s	30,000	kg
5	Moments of inertia	I_i	56,900	kgm^2
6	Rail car roll inertia	I_r	68,200	kgm^2
7	Rail car pitch inertia	I_p	71,000	kgm^2
8	Average mass of first wheel set and axle	m_1	1300	kg
9	Average mass of second wheel	m_2	1300	kg
10	Distance between the centre of gravity and the front position of the rail car	d_i	6	m
11	Distance between the centre of gravity and the middle position of the rail car	d_2	6	m
12	Distance between the centre of gravity and the rear position of the rail car	d_3	6	m
13	Spring constant of the primary suspension system	k_1	2.4×10^6	N/m
14	Spring constant of the secondary suspension system	k_2	5.6×10^5	N/m
15	Spring constant of the wheel	k_3	4.0×10^5	N/m
16	Damping constant of the primary suspension system	b_1	1.2×10^3	Ns/m
17	Damping constant of the secondary suspension system	b_2	2.95×10^4	Ns/m
18	Damping constant of the wheel	b_3	5.0×10^4	Ns/m

Source: [20, 21].

Table 1.
Input parameter for rail car system modeling.

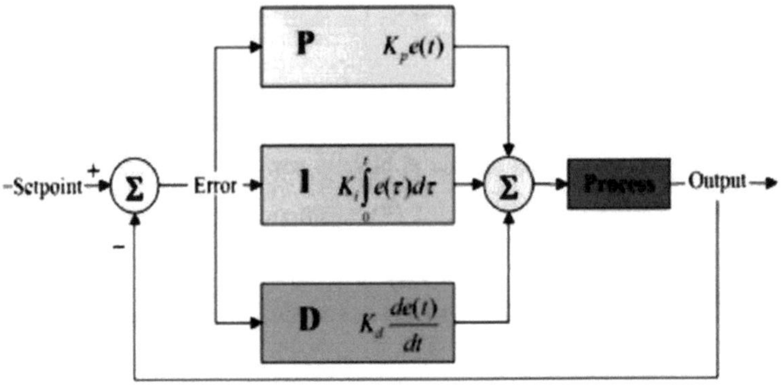

Figure 4.
The block diagram of the PID controller.

measurements and set points. The PID variables are iteratively adjusted until the steady-state error from the output signal is eliminated. This control action is done by adjusting the controller gain with resulting decrease in the rise time and percent increase in the overshoot, which makes the system go unstable. This rise time is further reduced with the integral control action. Finally, the derivative action is introduced to compensate for the offset. This reduces the percent overshoot and settling time, thus making the system stable over time.

Eq. (27) gives the expression for the control action of the PID controller:

$$u_c = K_p e(t) + \frac{K_i}{T_i}\int_0^t e(t)dt + K_d T_d \frac{de(t)}{dt} \tag{27}$$

The Nichols-Ziegler tuning rules employed for tuning the PID control as well as the summary of the effects of its control action on the PID are presented in **Tables 2** and **3**, respectively.

The signal (U) which passes through the controller computes the derivative and integral of error signal. The signal error is thereafter sent to the system in order to obtain the system's output (Y). The PID controller was designed in the MATLAB Simulink 2018 environment to generate a continuous time control. Using the Nichols-Ziegler rules, the tuning of PID controller was done by generating the

S/N	Type of controller	K_p	K_i	K_d
1	P	0.5 Kcr	∞	0
2	PI	0.45 Kcr	0.83 Pcr	0
3	PID	0.6 Kcr	0.5 Pcr	0.125 Pcr

Table 2.
Zeigler-Nichols tuning rules.

S/N	Controller response	Rise time	Overshoot	Setting time
1	K_p	Decrease	Increase	Small change
2	K_i	Decrease	Increase	Increase
3	K_d	Small change	Decrease	Decrease

Table 3.
Effect of the control action of the PID.

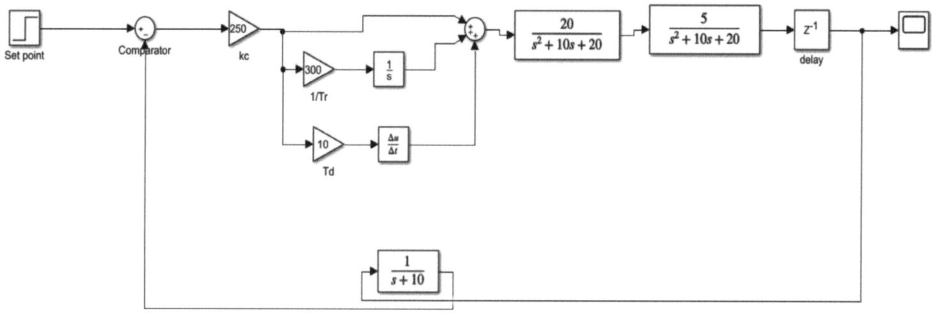

Figure 5.
The PID control system.

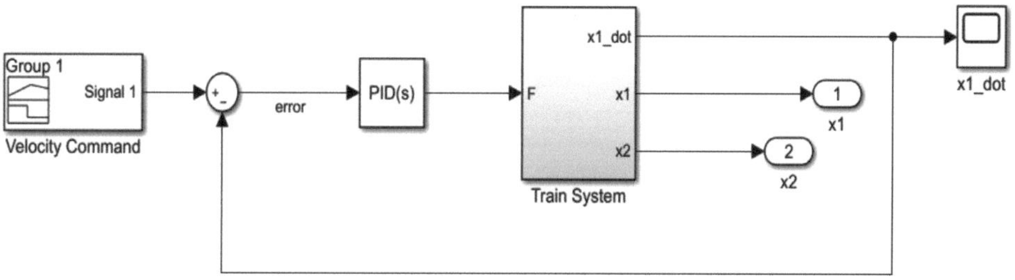

Figure 6.
The PID control and the rail car systems.

system's transfer function and subsequent importation of the parameters obtained into the linear time-invariant system.

The PID control system and its connection to the rail car system are shown in **Figures 5** and **6**, respectively.

The aim of the control design is to keep the system variables close to the reference in order to compensate for the effect of load and rail disturbances. The system requirement is to check unpleasant motion and ensure rail car stability by reaching a compromise between the stiff primary suspension and soft secondary suspension system.

The actively controlled suspension system can be activated via the use of solenoid, hydraulic, electromagnetic means or through a magnetorheological damper. This system is designed to use the solenoid actuators because of its lightweight, simplicity in structure, ease of installation and short response time, which makes it highly sensitive to disturbances.

3. Results and discussion

Figure 7 shows the step response before the iterative adjustment of the PID control. The amplitude of oscillation, which is a function of the percent overshoot, is 2 mm, and the system could not return to the equilibrium position after 3 s. The shape of the plot represents a system that is underdamped, which signifies the need for damping to minimize unwanted motion. The system whose step response is depicted in **Figure 7** is relatively unstable as vibration will reduce the system's and ride performance.

Figure 8 shows the step response from the controlled system. When compared to **Figure 1**, the amplitude of oscillation has reduced to 1.15 mm and settling time 0.5 s under the effect of the PID control action. The system is relatively stable as the

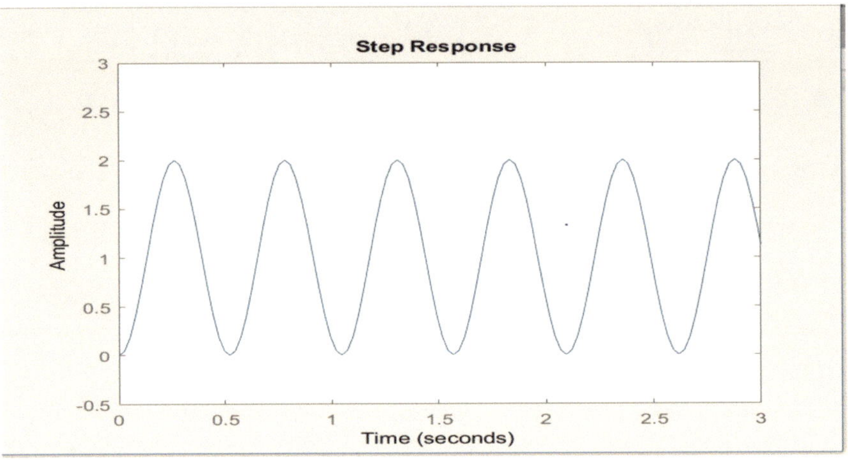

Figure 7.
Step response for the uncontrolled system.

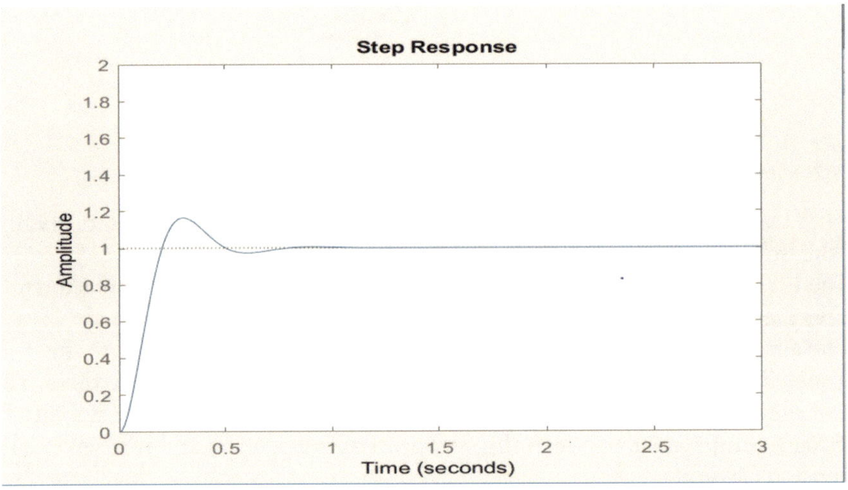

Figure 8.
Step response for the controlled system.

plot gives an indication of a critically damped system. This implies that under the effect of the PID control system, the system could regain its stability and equilibrium position after encountering some level of disturbances.

Figure 9 shows the response of the PID controller in order to determine whether the control system meets the design requirements. The tuned response represents the actual measurement of the process variables, while the baseline response represents the reference or set point. The differences between the tuned response and the baseline response give the steady-state error. The nature of the plot also indicates that the system is critically damped; thus, the damping ratio $\xi = 1$.

Figure 9 shows the reference tracking of the tuned response and the reference (baseline). The design requirement of the control system includes set-point tracking; hence, this plot shows how the closed-loop system responds to a step change in set point. From the plot, the steady-state error is minimal, thus indicating the effectiveness of the control system.

Figure 10 shows the closed-loop step response to a step disturbance at the system's output. This is important in analyzing the sensitivity of the control system

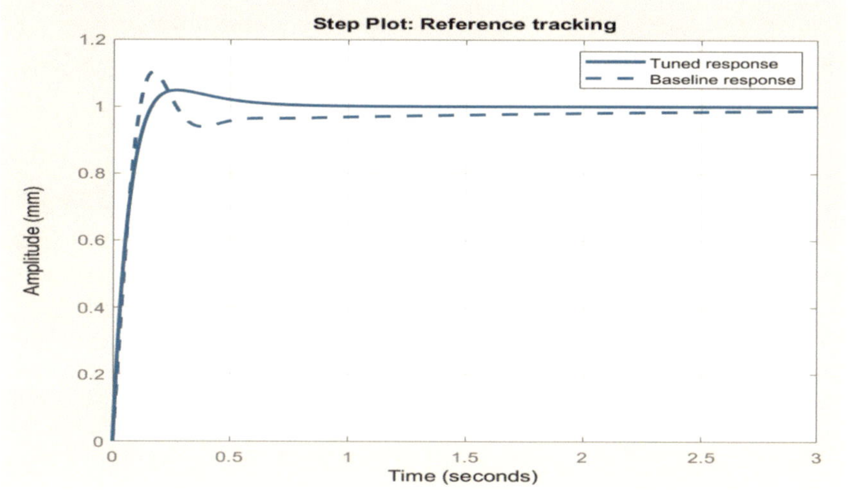

Figure 9.
Step plot for reference tracking.

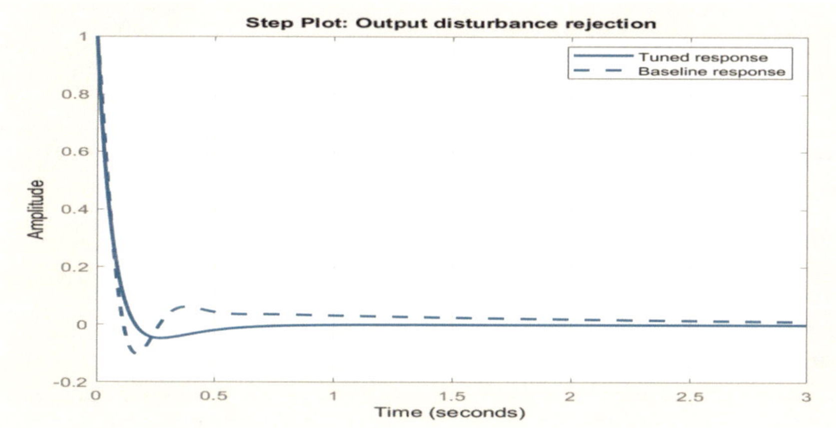

Figure 10.
Plot of output disturbance rejection.

to noise measurement. The negligible value of the steady-state error as shown by the degree of agreement between the tuned output response of the system and the baseline (reference) indicates the high sensitivity and compensation of the system's control to the noise measurement.

Figure 11 shows the closed-loop system response to load disturbance. The plot represents a step disturbance from the system's input. The differences between the tuned response and the baseline response, which is a function of the steady-state error, are significantly large. The implication of this is that the control system is insensitive to the rejection of load and other input disturbances, which is capable of offsetting the balance of the system. The large steady-state error resulting from input disturbance rejection stems from the fact that a single PID may not be able to satisfy all the design requirements at the same time; hence, there is always a performance trade-off amongst the reference tracking, percent overshoot and input disturbance rejection. However, the use of Fuzzy PID or ISA-PID controller can be used to meet the design requirements significantly. This will improve the response of the reference tracking with the provision of an additional tuning parameter,

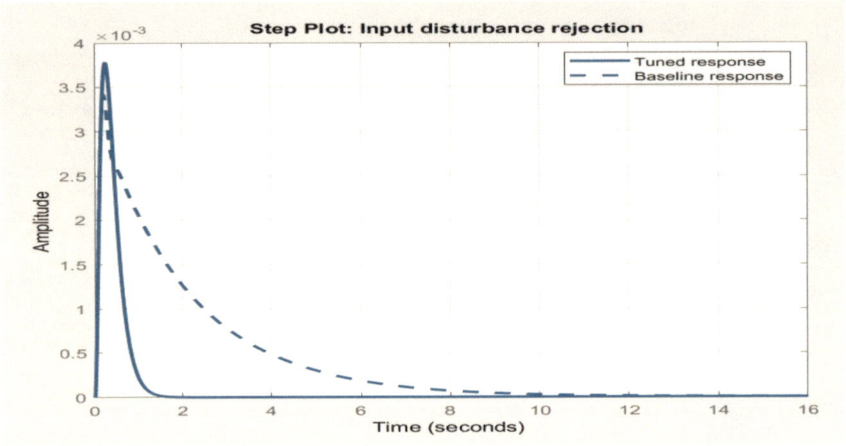

Figure 11.
Plot of input disturbance rejection.

Parameter	Tuned response	Baseline (reference) response
Rise time	0.07 s	0.0847 s
Settling time	0.499	1.94 s
Overshoot	4.76%	10.3%
Peak	1.05	1.1 mm
Phase margin	83°	60.4°
Frequency	16.6	17.8 rad
Closed loop	Stable	Stable

Table 4.
Summary of the performance of the PID control.

which allows for independent control of the effect of the reference signal on the proportional action.

The robustness and performance of the PID control are directly proportional to the degree of stability of the rail car, ride comfort and performance of the rail car. The summary of the performance of the PID control system is presented in **Table 4**.

From **Table 4**, the tuned response has better performance and robustness compared to the baseline due to periodic iterative adjustment to eliminate the steady-state error for each step input. The rise time of the tuned response indicates a fast response time of the control system to disturbances or changes when compared to the baseline. This signifies some degree of delay in the time it takes the system to respond to fluctuations. In addition, comparing the settling time for both responses, the tune response settles faster after some disturbances compared to the baseline response. This explains the ability of the rail car to regain its stability after encountering some level of disturbances with the active control system. Also, the percent overshoot for the tuned response is still within the range of the permissible oscillation (5%) for critically damped system which indicates that the rail car system is relatively stable amidst load and rail disturbances. An increase in the phase margins implies a significant reduction in the percent overshoot and bandwidth.

Figure 12 shows the bode response of the control system. This is the plot of the frequency and phase response of the control system. The phase margin was found to

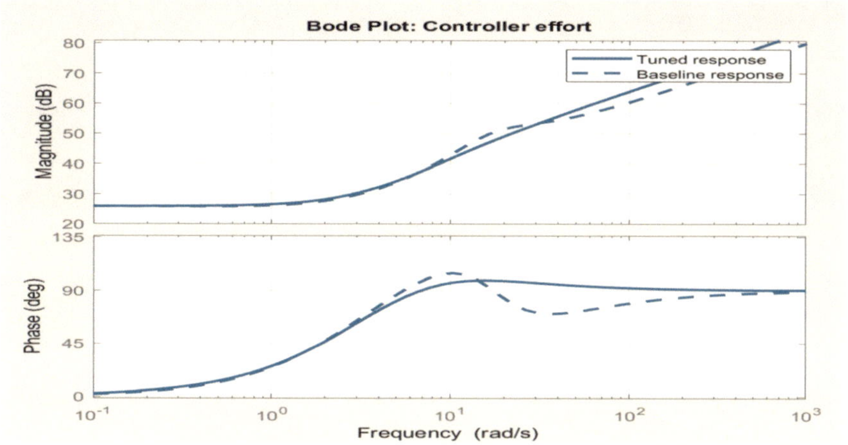

Figure 12.
Bode plot for the controller effort.

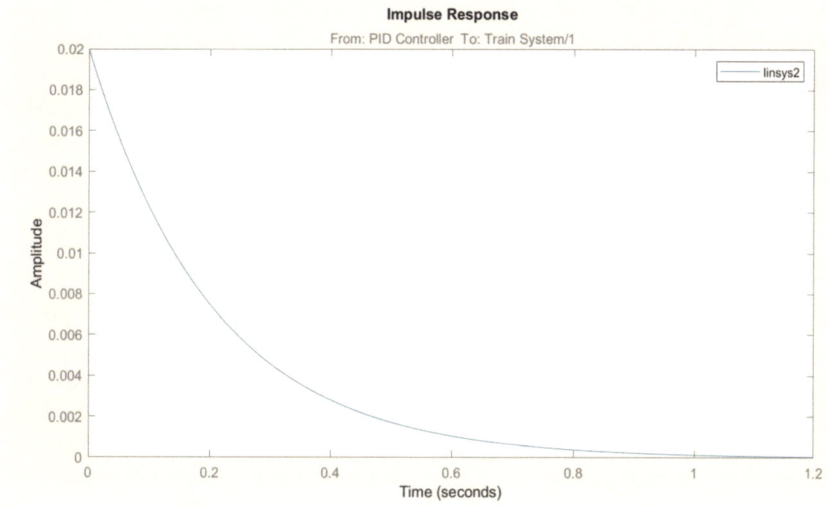

Figure 13.
Impulse response of the actively controlled rail car.

be 83° at a frequency of 16.6 rad/s for the tuned response compared to the baseline response with a phase margin and frequency of 60.4° and 17.8 rad/s, respectively. The relationship between the phase margin and percent overshoot is inversely proportional. Hence, the high value of the phase margin results in significant decrease in the percent overshoot, thus improving the rail car stability. Also, the higher closed-loop bandwidth results in faster rise time. The rise time was found to be 0.07 s for the tuned response and 0.0847 s for the baseline response. This implies that for the baseline response, the percent overshoot is still significant to offset the stability of the rail car system.

Figures 13 and **14** show the result of the linearization of the rail car system. This is to determine the dynamics of the system in real time and within time and frequency domains. **Figure 13** shows the impulse response of the rail car system, which is the degree of rail car body displacement as a result of load or rail disturbances. The maximum amplitude oscillation is 0.02 mm, which is negligible and insufficient to offset the rail car balance. In addition, the settling time (less than 1 s)

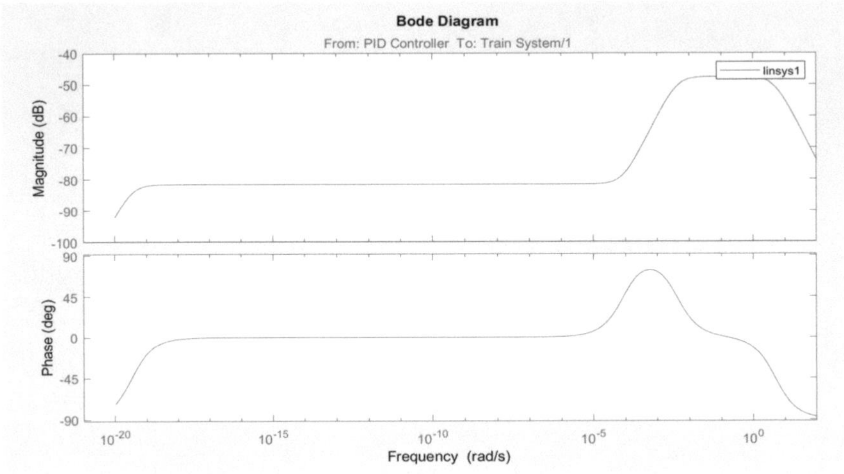

Figure 14.
Bode plot for the actively controlled rail car.

is also within the permissible range indicating the ability of the system to regain its stability within a second as a result of load or rail disturbances.

Figure 14 shows the bode plot of the rail car system; the phase margin was found to be 48° at a frequency of 20.2 rad/s which signifies significant reduction in the percent overshoot due to the compensation for steady-state error by the derivative action of the PID.

4. Conclusion

The control of undesirable vibrations of the rail car system in 6 degrees of freedom was carried out using the PID control system. The initial displacement and vertical acceleration that characterize the performance of the system on encountering rail and load disturbances were minimized with the iterative adjustment of the PID controller according to Nichols-Ziegler tuning rules. In addition, the performance of the control and the rail car system in terms of the input step response, bandwidth, frequency, phase margin and input and output rejections was within the acceptable range. Hence, the PID control system shows significant robustness in providing the required active control for the system, while the rail car system shows improved stability and reduction in vibration under control action of the PID, thus improving ride comfort. However, a single PID may not sufficiently satisfy all the design requirements at the same time resulting in performance trade-off. However, Fuzzy PID or ISA-PID controller can be used to meet the design requirements significantly. This will further improve the performance and robustness of the rail car system.

DOI: http://dx.doi.org/10.5772/intechopen.85654

Author details

Ilesanmi Afolabi Daniyan* and Khumbulani Mpofu
Department of Industrial Engineering, Tshwane University of Technology, Pretoria, South Africa

*Address all correspondence to: afolabiilesanmi@yahoo.com

References

[1] Al-Zughaibi A, Davies H. Controller design for active suspension system of quarter car with unknown mass and time-delay. International Journal of Mechanical, Aerospace, Industrial, Mechatronic and Manufacturing Engineering. 2015;**9**(8):1484-1489

[2] Sharma RC, Sharma SK. Sensitivity analysis of three-wheel vehicle's suspension parameters influencing ride behaviour. Noise & Vibration Worldwide. 2018;**49**(7–8):272-280

[3] Bideleh SMM, Mei TX, Berbyuk V. Robust control and actuator dynamics compensation for railway vehicles. Vehicle System Dynamics. 2016;**54**(12): 1762-1784

[4] Yao JL, Shi WK. Development of a sliding mode controller for semi-active vehicle suspensions. Journal of Vibration and Control. 2013;**19**: 1152-1160

[5] Arefsoliman MA. Adaptive LQR control strategy for active suspension system. 2011. SAE Technical Paper, 2011-01-0430

[6] Herbst G. A simulative study on active disturbance rejection control (ADRC) as a control tool for practitioners. Electronics. 2013;**2**(3): 246-279

[7] Colombo EF, Gialleonardo ED, Facchinetti A, Bruni S. Active car body roll control in railway vehicles using hydraulic actuation. Control Engineering Practice. 2014;**31**:24-34

[8] Gohrle C, Schindler A, Wagner A, Sawodny O. Design and vehicle implementation of preview active suspension controllers. IEEE Transactions on Control Systems Technology. 2014;**22**(3): 1135-1142

[9] Jin X-S, Xiao X-B, Ling L, Zhou L, Xiong J-Y. Study on safety boundary for high-speed trains running in severe environments. International Journal of Rail Transportation. 2013;**1**(1–2):87-108. DOI: 10.1080/23248378.2013.790138

[10] Nguyen SD, Choi SB, Nguyen QH. An optimal design of interval type-2 fuzzy logic system with various experiments including magnetorheological fluid damper. IMechE Part C: Journal of Mechanical Engineering Science. 2014;**228**(17): 3090-3106

[11] Nguyen SD, Nguyen QH. Design of active suspension controller for train cars based on sliding mode control, uncertainty observer and neuro-fuzzy system. Journal of Vibration and Control. 2017;**23**(8):1334-1353

[12] Kim HC, Choi SB, Lee GS. Performance analysis of a semi-active railway vehicle suspension featuring MR dampers. In: SPIE Smart Structures and Materials + Non-destructive Evaluation and Health Monitoring, San Diego, CA. Bellingham, WA: International Society for Optics and Photonics (SPIE); 2014. p. 905711

[13] Kumar PS, Sivakumar K, Kanagarajan R, Kuberan S. 2799 adaptive neuro fuzzy inference system control of active suspension system with actuator dynamics. Journal of Vibroengineering. 2018;**20**(1):541-549

[14] Matamoros-Sanchez AZ, Goodall RM. Novel mechatronic solutions incorporating inerters for railway vehicle vertical secondary suspensions. Vehicle System Dynamics. 2015;**53**(2): 13-136

[15] Oh J-S, Shin Y-J, Koo H-W, Kim H-C, Park J, Choi S-B. Vibration control of a semi-active railway vehicle

suspension with magneto-rheological dampers. Advances in Mechanical Engineering. 2016;**8**(4):1-13

[16] Suarez B, Felez J, Maroto J, Rodriguez P. Sensitivity analysis to assess the influence of the inertial properties of railway vehicle bodies on the vehicle's dynamic behaviour. Vehicle System Dynamics. 2013;**51**(2): 251-279

[17] Zolotas AG, Goodall RM. Modelling and control of railway vehicle suspensions. In: Turner MC, Bates DG, editors. Mathematical Methods for Robust and Nonlinear Control, Lecture Notes in Control and Information Sciences. New York: Springer; 2007. pp. 373-412

[18] Liang L, Xiao X-B, Xiong J-Y, Zhou L, Wen Z-F, Jin X-S. A three-dimensional model for coupling dynamics analysis of high speed train-track system. Journal of Zhejiang University Science A (Applied Physics and Engineering). 2014:1-21

[19] Sezer S, Atalay AE. Application of fuzzy logic based control algorithms on a railway vehicle considering random track irregularities. Journal of Vibration and Control. 2011;**18**(8):1177-1198

[20] Ulum Z, Affaf M, Salmah JCE, Suparwanto A. Active suspension systems design of a light rail vehicle using MPC with preview information disturbance. In: 5th International Conference on Instrumentation, Control, and Automation (ICA) Yogyakarta, Indonesia, August 9–11, 2017. pp. 18-23

[21] Sharma SK, Kumar A. Ride performance of a high-speed rail vehicle using controlled semi active suspension system. Smart Materials and Structures. 2017;**26**(1–19):055026

Chapter 8

Analysing Non-Linear Flutter Vibrations Using System Dynamic Approach

Cosmas Pandit Pagwiwoko and Louis Jezequel

Abstract

The objective of this work is to investigate the dynamic behaviour of aero-elastic vibrations in the presence of non-linear stiffness such as free-play mechanism, softening or hardening stiffness. A closed loop dynamic system is proposed to represent the phenomenon of flow-structure interaction. In this approach a transfer function for generating the aerodynamic forces based on the structural response is constructed in the feedback loop of the dynamic system with the aid of Padé rational function. The effects of the non-linear factors therefore can be included conveniently in time domain simulation and the stability of limit cycle oscillations (LCO) can be analysed accurately.

Keywords: LCO, non-linear structures, flow structure interactions, aero-elasticity, self-excited vibration, binary classical flutter

1. Introduction

Aero-elasticity is a multi-physics discipline that involves the loads of aerodynamics, elastic and inertial generated by the motion of structure. One of the most important phenomena in this field is flutter regarding to its harmful effect to the structure. This flow-induced vibration under certain conditions can be self-excited and divergently unstable. In aerospace industry the boundary of flutter instability is usually determined by V-g method, a computational technique in frequency domain based on the balance of energy of the oscillating wing and the flow by maintaining a harmonic function of the aero-elastic response. In this method explained remarkably by Stanciu et al. [1], the critical velocity of the flow is determined by solving the complex eigen value problem of the aero-elastic system. Although the method can give accurate results it is only effective for linear cases. However in real aircraft there are non-linear factors of structure such as free-play, hysteretic and large deformation that need to be taken into account. Trickey et al. [2] observed that certain cases in regards to excessive LCO found in some Boeing and Airbus aircrafts and stated that the characterization and explanation of this non-linear vibration were important for fatigue and maintenance issues. In their research, the methods of non-linear dynamics were developed for these purposes and they proposed a novel system-identification technique to generate an approximation of LCO to be used for online monitoring of dynamic behaviour close to bifurcation condition [8].

Pereira et al. [3] showed an example of LCO due to the existence of hardening nonlinearity of wing stiffness in pitching of F-16 aircraft that caused persistent aero-elastic problems. Therefore the knowledge and comprehension of non-linear aero-elasticity are of increasing importance in aircraft design. In their work an investigation on the combined influence of hardening and free-play nonlinearities on the bifurcation response was carried out.

A document regarding the missing of MH370 Boeing 777 is also concerned about the phenomena of aero-elasticity. The failure analysis of the right-side flaperon that was found in French territory's Reunion Island [4] on 2015, reported that flutter (LCO) caused to repetitive loading which in turn imparted stress fatigue in the primary aluminum alloy attachment components.

This work has an intention to simulate numerically the interaction of flow-structure as a dynamic system by arranging the structure part as a principle plant and the aerodynamic part as a feedback loop subsystem. The analysis of structural response in time domain enables to insert the nonlinearities conveniently. The part of structures is reconstructed in a form of block-diagram representing a dynamic system where the inertial loads are expressed explicitly as a result of the elastic loads and frictions generated in the progressing structure response due to external aero-dynamic excitations, while the part of aerodynamics is arranged as a feedback-loop transfer function activated by the structural response. For this purpose, the unsteady aerodynamic forces calculated by using singularity method in frequency domain have to be converted to Laplace variable s by using Padé's approximation rational function. Botez et al. [5] in conducting flutter analysis of CL-604 Bombardier, used a least-squares technique utilizing certain number of lagging-terms, and the approximation showed the best aerodynamic forces conversion from frequency into Laplace domain in terms of execution time and precision.

In analysing LCO on mechanical system in general where there is involvement of various physical parameters such as non-linear stiffness, hysteretic and free-play, and more specifically the influence of damping to the stability of the oscillations, Sinou and Jézéquel [6] proposed to employ a two-degree-of-freedom model for the sake of simplicity. With the same spirit, in this study we use a pitch-plunge two dimensional wing-section model in analysing the effects of structural nonlinearities on a binary classical flutter.

2. Stability analysis of aero-elastic system

2.1 Description of the two-degree-of-freedom model

Figure 1 shows schematically a two-degree-of-freedom pitch-plunge fluttering aerofoil model. The support system consisting of axial and rotational springs are attached to a rigid aerofoil on a point so-called elastic axis. These two flexible supports restrict the motions of the aerofoil with the exception in the two modes of translation and rotation. For the case of zero damping, the equations of motion of the aerofoil subjected a uniform flow can be written as:

$$\begin{bmatrix} m & mx_\alpha b \\ mx_\alpha b & I_g + m(x_\alpha b)^2 \end{bmatrix} \begin{Bmatrix} \ddot{h} \\ \ddot{\alpha} \end{Bmatrix} + \begin{bmatrix} k_h & 0 \\ 0 & k_\alpha \end{bmatrix} \begin{Bmatrix} h \\ \alpha \end{Bmatrix} = \begin{Bmatrix} -L \\ Lec + M_{ac} \end{Bmatrix} = \{F_{aero}\} \tag{1}$$

where h and α are the degree of freedom in plunging and pitching, respectively, as explained in **Figure 1**, while k_h and k_α are the spring stiffness in translation and rotation, respectively.

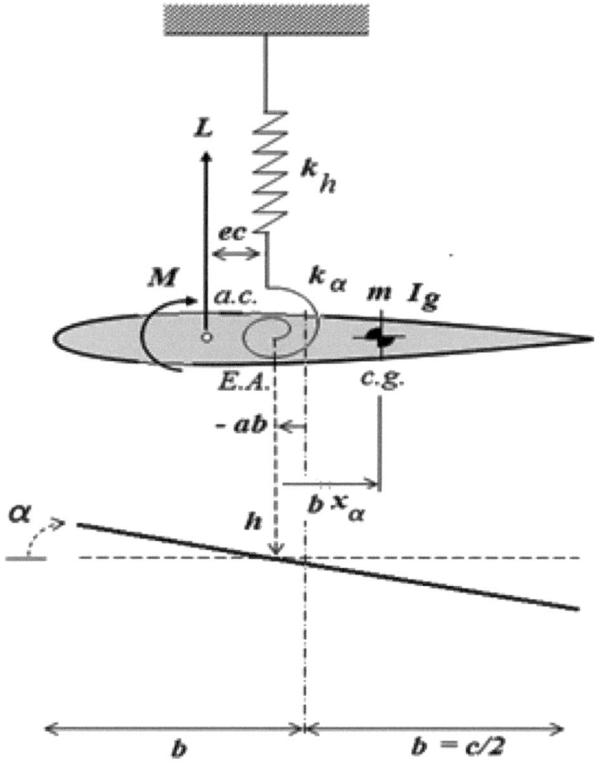

Figure 1.
Two-degree-of-freedom aero-elastic model.

2.2 Unsteady aerodynamic model

Theodorsen's unsteady aerodynamic model as explained by Brunton and Rowly [7] excellently is used in this work. This method analyses the motions of the aerofoil in frequency domain and it assumes that the amplitudes are small. In the analysis the aerofoil is considered thin, the flow is inviscid incompressible with no separation or intrusion.

In this method the frequencies of the harmonic oscillating motions are considered relatively slow therefore the transversal and rotational velocities of the aerofoil contribute as an additional angle of attack to the total lift. As a result the quasi-steady of the lift coefficient C_L^{QS} can be expressed proportional to the total angle of attack:

$$C_L^{QS} = \frac{\partial C_l}{\partial \alpha}\left(\alpha + \frac{\dot{h}}{U_\infty} + b\left(\frac{1}{2} - a\right)\frac{\dot{\alpha}}{U_\infty}\right) \quad (2)$$

In thin aerofoil theory the lift gradient can be considered equals to 2π, the vortex singularity is located at the aerodynamic center (a quarter of the chord from the leading-edge) and the downwash velocity is focused at three quarter of the chord.

The aerodynamic loading consisted of lift and pitching moment can be presented as:

$$L = \frac{1}{2}\rho U^2 c l \, C(k) \, C_L^{QS} \quad (3)$$

$$M_{ac} = \frac{1}{2}\rho U^2 c^2 l \, C(k) \, C_L^{QS} \quad (4)$$

where c and l are the chord and the span of the 2D wing, respectively. $C(k)$ is Theodorsen's function showing that there is a phase difference between aerodynamic loading and wing section's motion. The parameter k called reduced (nondimensional) frequency is defined as $\omega b/U$.

2.3 Flutter stability boundary

V-g method based on the balance of energy between the flow and the motion of the structure is used to determine the flutter boundary for a linear aero-elastic system. The analysis is conducted in frequency domain where harmonic motions in pitching and plunging are imposed to the dynamic response of the structure to represent the state of the aero-elastic system in the stability boundary. In order to maintain harmonic motions of the structure, a virtual structural damping g is inserted to the system hence the equation of motions adopting from Eq. (1) becomes:

$$\begin{bmatrix} m & mx_\alpha b \\ mx_\alpha b & I_g + m(x_\alpha b)^2 \end{bmatrix} \begin{Bmatrix} \ddot{h} \\ \ddot{\alpha} \end{Bmatrix} + (1+ig) \begin{bmatrix} k_h & 0 \\ 0 & k_\alpha \end{bmatrix} \begin{Bmatrix} h \\ \alpha \end{Bmatrix} = \{F_{aero}\} \tag{5}$$

By imposing a harmonic functions to the motions in both plunging and pitching as shown in Eq. (6):

$$\begin{Bmatrix} h \\ \alpha \end{Bmatrix} = \begin{Bmatrix} \hat{h} \\ \hat{\alpha} \end{Bmatrix} e^{i\omega t} \tag{6}$$

Henceforth, the aerodynamic loading can be represented as the expression in Eq. (7):

$$\{F_{aero}\} = \begin{Bmatrix} -L \\ Lec + M_{ac} \end{Bmatrix} = \frac{1}{2}\rho U^2 [Q(ik)] \begin{Bmatrix} h \\ \alpha \end{Bmatrix} \tag{7}$$

where the generalized aerodynamic matrix $Q(ik)$ in complex form can be written as:

$$[Q(ik)] = cl\,C(k) \times \left(\begin{bmatrix} 0 & \left(-\frac{\partial C_l}{\partial \alpha}\right) \\ 0 & \left(2\frac{\partial C_l}{\partial \alpha} be + 2\frac{\partial C_m}{\partial \alpha} b\right) \end{bmatrix} + i \begin{bmatrix} \left(-\frac{\partial C_l}{\partial \alpha}\frac{k}{b}\right) & 0 \\ \left(2ek\frac{\partial C_l}{\partial \alpha} + 2k\frac{\partial C_m}{\partial \alpha}\right) & 0 \end{bmatrix} \right) \tag{8}$$

subsequently the equations of motion showed in Eq. (5) will lead to the solution of the eigen values problem as presented below:

$$\begin{bmatrix} k_h & 0 \\ 0 & k_\alpha \end{bmatrix}^{-1} \left(\begin{bmatrix} m & mx_\alpha b \\ mx_\alpha b & I_g + m(x_\alpha b)^2 \end{bmatrix} + \frac{\rho b^2}{2k^2}[Q(ik)] \right) \begin{Bmatrix} \hat{h} \\ \hat{\alpha} \end{Bmatrix} = \frac{(1+ig)}{\omega^2} \begin{Bmatrix} \hat{h} \\ \hat{\alpha} \end{Bmatrix} \tag{9}$$

Elaborating a certain range of reduced frequency k into Eq. (9) will give as a result, a range of complex eigenvalues for both modes of motion. The real parts relate to the natural frequencies while the imaginary parts to the artificial structural damping of the aero-elastic system for certain values of flow velocity, associated with the reduced frequency k (**Table 1**).

An aero-elastic flutter model with NACA0015 wing section is fabricated and installed in the wind-tunnel in vertical position as shown in **Figure 2a** and **b**. The each end of the wing model is mounted on a support system consisted of a pair of steel cantilever beams to allow the side-slipping translation motion and a warm spring for the yawing rotational motion.

The solution of the eigenvalue problem in Eq. (9) yields two curves of natural frequencies and two artificial structural damping depending to the flow speed as presented in **Figures 3** and **4**, respectively. The critical speed is defined where one of the artificial structural damping curves intercepts the real actual structural damping of the structure.

Figure 4 shows the flutter boundary of this linear aero-elastic model where the critical flow speed is around 15 m/s. The negative values of the damping curves indicate that an amount of energy has to be supplied to the system to have a

Notation	Description	Value
a	Relative distance to half chord from midchord	−0.5
b	Half chord	0.05 m
c	Chord	0.1 m
l	Span of the wing model	0.28 m
e	Relative distance from $E.A.$ to $a.c.$	0.0
x_α	Relative distance from $c.g.$ to mid-chord in b	0.34
m	Plunging mass	1.35 kg
m_p	Pitching mass	0.93 kg
I_p	Mass moment of inertia in pitching	$7.741e^{-4}$ kg m^2
k_h	Axial stiffness in plunging	906.81 N/m
k_α	Rotational stiffness in pitching	11,445 Nm
$E.A.$	Elastic axis	
$a.c.$	Aerodynamic center	
$c.g.$	Centre of gravity	

Table 1.
Value of physical parameters.

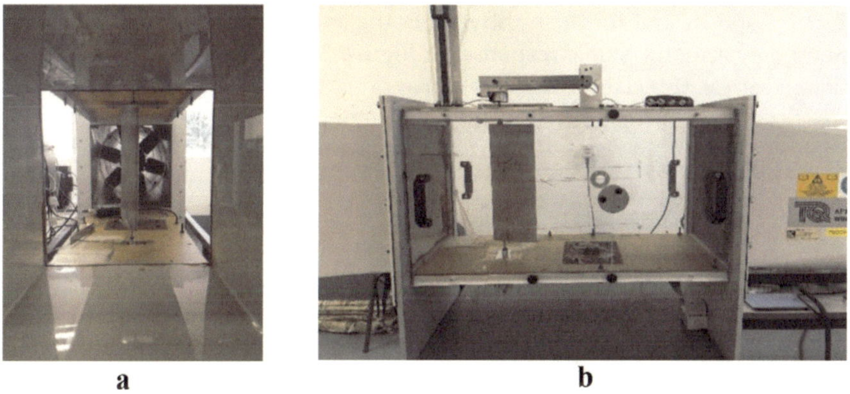

Figure 2.
Aero-elastic wind-tunnel model, NACA 0015 aerofoil, two-degree-of-freedom system in translation and rotation, in the cross-section of 30 cm × 30 cm test-section.

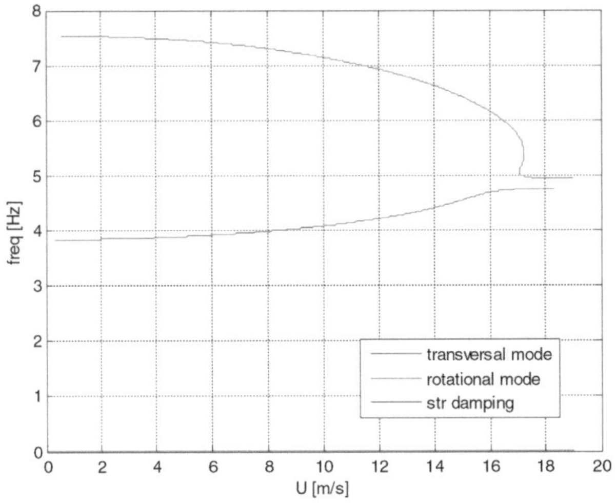

Figure 3.
Natural frequencies in translation and rotation modes (in Hertz) versus flow velocity (in m/s).

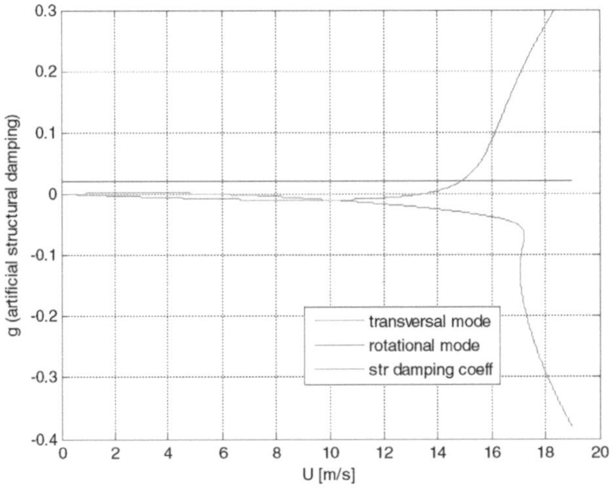

Figure 4.
Artificial structural damping parameters g in translation and rotation modes versus flow velocity (in m/s). The solid horizontal red colour line is the value of structural damping of steel (the material of the both springs).

harmonic response and for the positive ones the energy has to be dissipated such to maintain the harmonic stable response. In **Figure 3**, the frequencies of two aero-elastic modes, i.e. translation and rotation approach each other as the flow speed is getting close to the critical speed of flutter boundary. The phenomenon so-called internal resonance shows that there is an interchange of energy between the two modes of vibration.

3. Non-linear behaviour of the aero-elastic system

3.1 System dynamic approach

Consider a structure with M, C, K as the matrices of mass, damping and stiffness respectively, subjected to an external loads $F(t)$. The dynamic response $x(t)$ basically can be presented in an arrangement of block diagrams based on the equation of motion by showing explicitly the inertial internal loads:

$$M\ddot{x} = F(t) - C\dot{x} - Kx \tag{10}$$

The above expression can be considered as a junction with the input of $F(t)$ which consists of a disturbance and aerodynamic forces encountered by closed loop feedback signals of C and K, brings forth the output of inertial load signals. **Figure 5** explains the block diagram of this mechanical system.

To be able to model the effects of structural non-linearity accurately and to simulate conveniently in time domain, the aero-elastic system needs to be

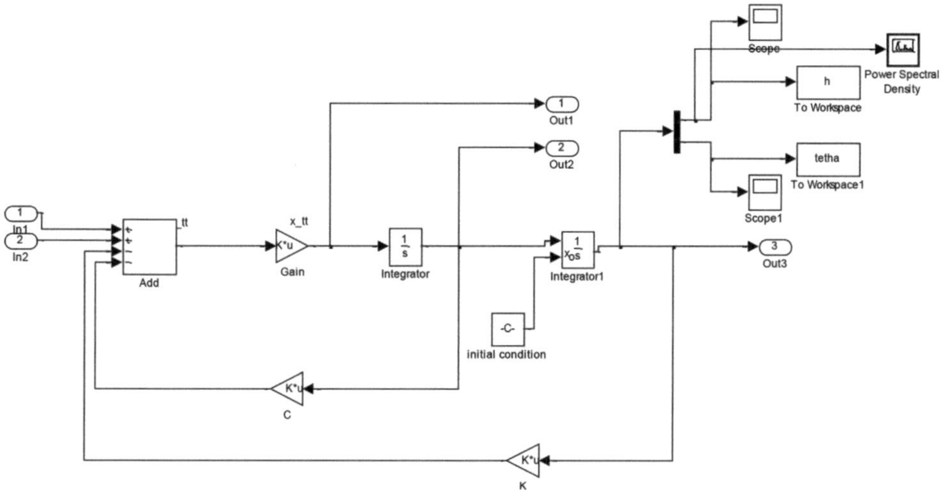

Figure 5.
Block diagram for simulation of a structure subjected to an external load and some initial conditions.

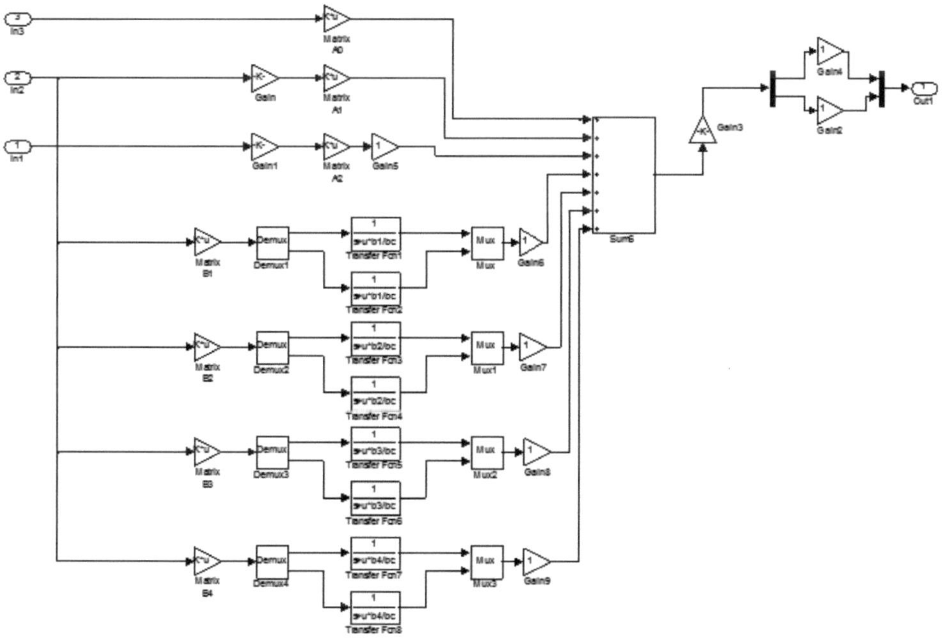

Figure 6.
(A) Matrix Q11 signifying the generalized aerodynamic forces where the points are values of calculated in frequency domain for a certain range of reduced frequency k*, and the solid line is the approximated curve in Laplace variable s by using four parameters of lagging term* β_j *of 01, 0.2, 0.3 and 0.4. (B) Matrix Q21 signifying the generalized aerodynamic forces where the points are values of calculated in frequency domain for a certain range of reduced frequency* k*, and the solid line is the approximated curve in Laplace variable s by using four parameters of lagging term* β_j *of 01, 0.2, 0.3 and 0.4.*

represented as a system dynamic model. In this system the aerodynamic forces are generated by the lifting surface as a result of the structural temporal response in a closed loop form. For this purpose the unsteady-aerodynamic forces calculated in frequency domain based on harmonic motions of the natural modes as expressed in Eq. (8), are now converted in Laplace variable using Padé rational function approximation as shown in Eq. (11).

The lagging term parameters β_j, for $j = 1, n$, are real and chosen less than 1, where the values and the numbers are determined to optimize the approximated curves. The matrices A_0, A_1 and A_2 in real values are estimated with curve-fitting using least square technique in complex plan to approximate the values of the aerodynamic forces calculated in frequency domain.

$$[Q(ik)] \approx [A_0] + [A_1]\left(\frac{bs}{U_\infty}\right) + [A_2]\left(\frac{bs}{U_\infty}\right)^2 + \sum_{j=1}^{n} \frac{[A_{j+2}]s}{s + \beta_j \frac{U_\infty}{b}} \tag{11}$$

Figure 6A and **B** explains the curve fitting of the generalised aerodynamic forces in matrix Q11 and Q21 calculated in a range of frequency in discretized data with the approximation rational function of Padé showed in solid lines.

The other matrix of the generalised aerodynamic forces Q12 and Q22 are zero as the consequence of the location of the elastic axis coincides with the aerodynamic center of the wing section.

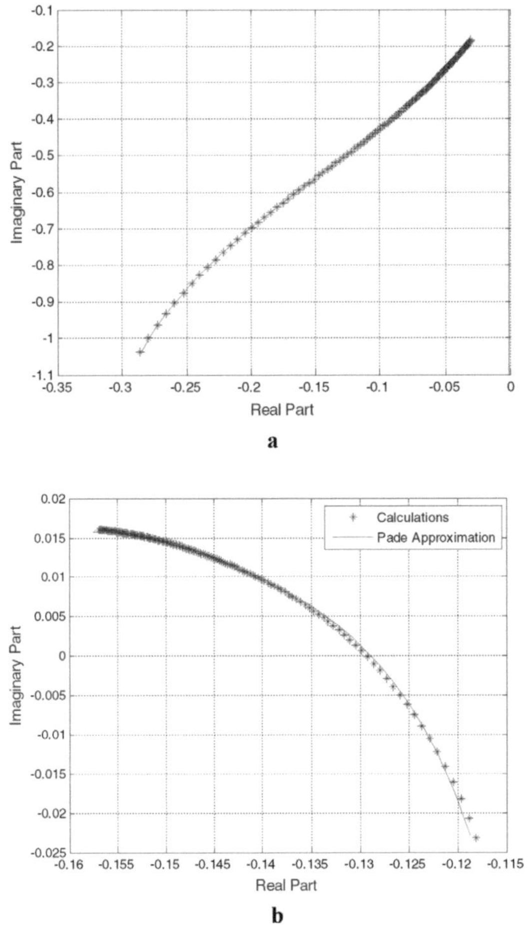

Figure 7.
Aerodynamic transfer function.

Based on the dynamic response of structure in terms of $x(t)$, $d/dt\,x(t)$ and d2/dt2 x(t) the aerodynamic forces in the forms of lift and pitching moment can be calculated by using a transfer function constructed using Eq. (10). **Figure 7** shows the arrangement of the block diagrams to express the aerodynamic transfer function.

The interaction of flow and structure can be simulated by arranging the block diagrams of structure showed in **Figure 5** as a subsystem, coupled with the aerodynamic forces showed in **Figure 7** where the calculation is conducted in time domain for each step of time discrete, simultaneously. The block diagrams of the plant representing the structure and the aerodynamic forces subsystem as a feedback loop have to be arranged such that all the processes are enhanced in integral operations to ensure the minimum numerical errors and the convergence of the solutions.

By putting together the block diagram representing the structure as shown in **Figure 5** with the aerodynamic transfer function shown in **Figure 7** as a feedback loop based on the structural response to generated aerodynamic forces, the flow structure interaction can be represented as two subsystems interconnected to each other, triggered by a disturbance subsystem as explained in **Figure 8**.

For validating the numerical model and simulation showed in **Figure 8**, the case of linear elastic of the aero-elastic system is conducted first. **Figures 9** and **10** shows

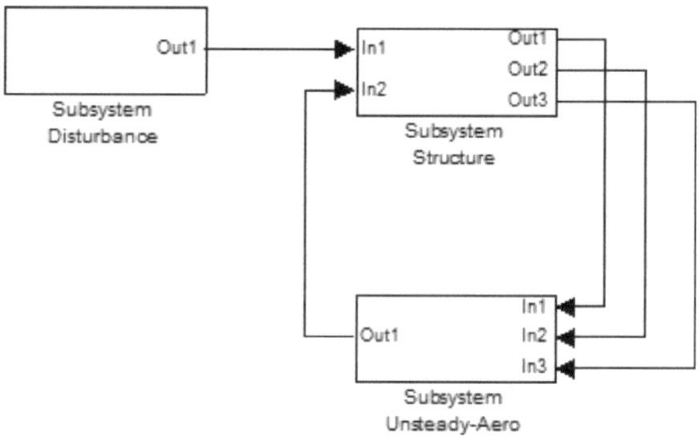

Figure 8.
Simulation of flow-structure interactions is carried out by arranging the subsystem of structure containing mass, mass moment of inertia, structural stiffness and damping, coupled with aerodynamic forces in lift and pitching moment as functions of the dynamic response of structure.

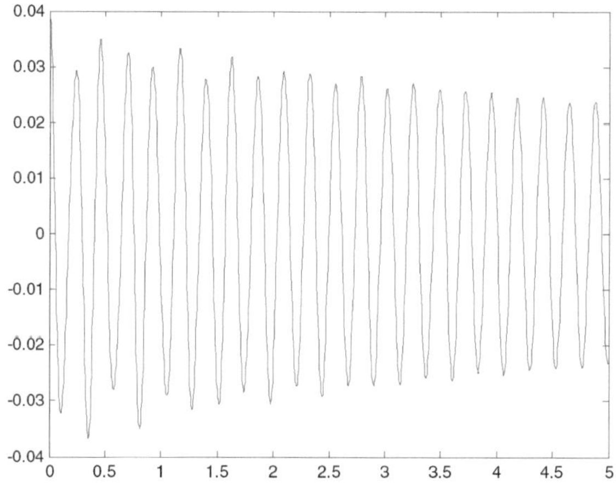

Figure 9.
Convergent translation response of the aerofoil at the flow speed of 13 m/s with initial condition of 4 cm displacement.

the response of the aerofoil in translation for the airflow speeds of below and above the flutter boundary.

The simulation for the aero-elastic system at critical airflow under a certain disturbance or initial condition will evidently yield a constant amplitude of sinusoidal motions.

3.2 Flutter limit cycle oscillation

Phase portraits representing the relationships between the displacement and the velocity of the response are used to analyse the dynamic behaviour of the non-linear system.

Figure 11 presents the phase diagram or the case of linear aero-elastic system where the ellipsoidal trajectories show a stable harmonic response of flutter boundary at the wind speed of flutter boundary.

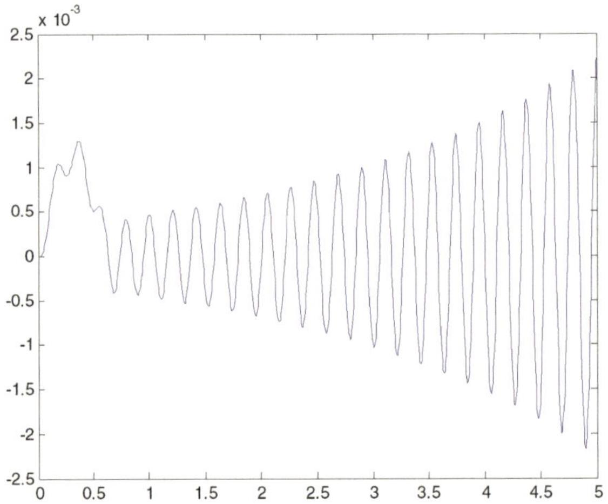

Figure 10.
Divergent unstable response of the aerofoil due to a small impulse disturbance at the flow speed of 16 m/s in translation.

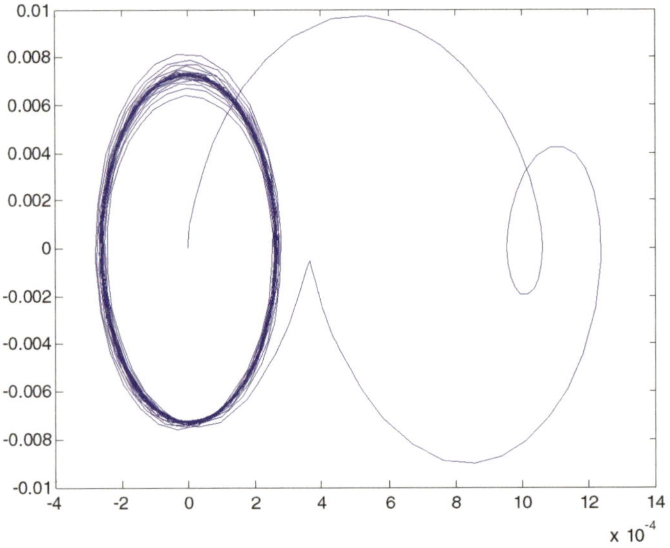

Figure 11.
Phase portrait of the linear aero-elastic system at the flutter speed boundary of 14 m/s.

A structural non-linear factor may influence the dynamic behaviour of an aero-elastic system. The insertion a free-play in rotation into the support mechanism of the model for an example, at the wind speed below the critical flutter, the system will not reduce entirely the dynamic response of the aerofoil under a certain perturbation as in the linear case, but introduce an oscillation with constant amplitude at a certain frequency. **Figure 12** shows a limit cycle oscillation at the wind speed of 13.0 m/s of the aero-elastic system under an initial condition.

Furthermore structural non-linear factors may also influence the limit of stability. The free-play mechanism reduces the flutter critical speed for around 0.5 m/s as showed in **Figures 13** and **14**.

At the critical speed calculated for linear system but with the existence of the free-play, the system will generate an unstable divergent structural response as explained in **Figure 14**.

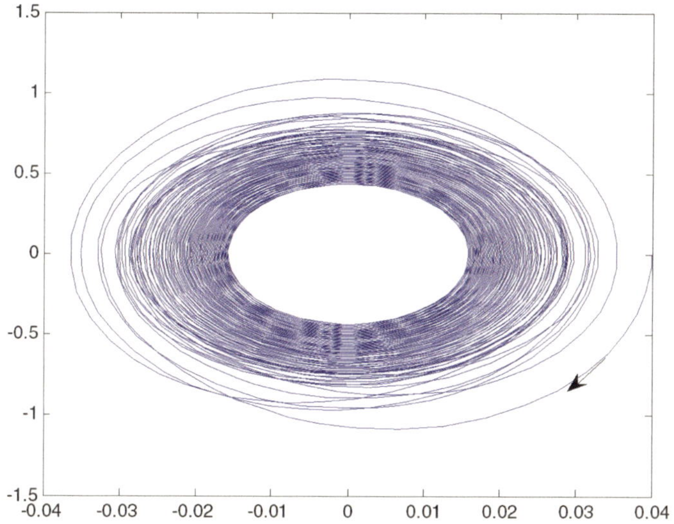

Figure 12.
Phase portrait of the aero-elastic system with free-play mechanism of 2° of rotation at the wind speed of 13.0 m/s due to initial displacement of 4 cm.

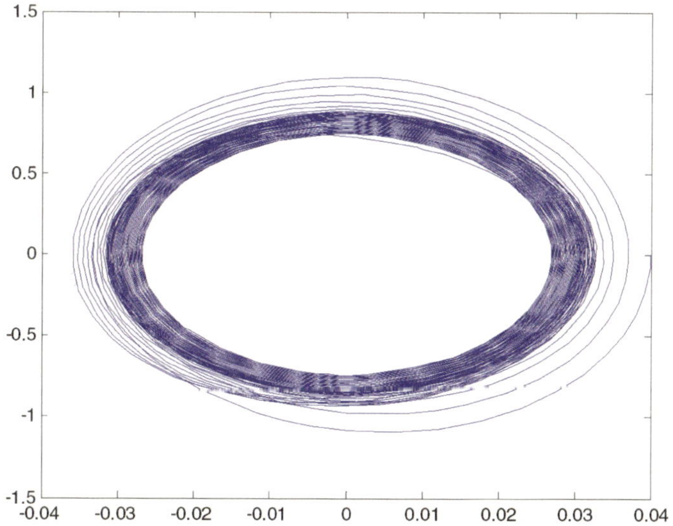

Figure 13.
Phase portrait for the existence of 2° free-play mechanism in rotation at the wind speed of 13.5 m/s with the initial condition of 4 cm displacement.

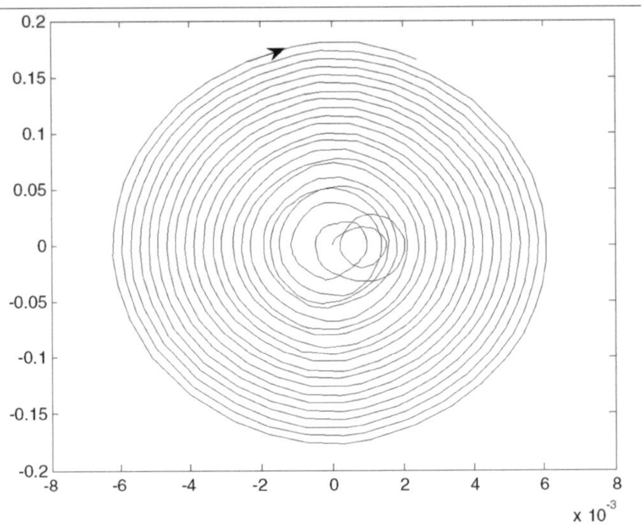

Figure 14.
Phase portrait for the existence of 2° free-play mechanism in rotation at the wind speed of 14.0 m/s.

From the wind-tunnel flutter testing it is observed that a moderate oscillation starts at 13 m/s flow speed and becomes severe vibration at 18 m/s. It can be concluded that a small free-play mechanism involves in the lower speed (less than the critical boundary), and a hardening-stiffness behaviour for the higher speed.

4. Conclusion

A two-degree-of-freedom in transversal and rotational motions wing section for describing classical binary flutter mechanism is used to investigate the effect of free-play nonlinearity to the stability of the aero-elastic system and the associated limit cycles. The aerodynamic forces are calculated by using Theodorsen's method in frequency domain based on thin aerofoil theory.

By representing the aero-elastic system as a closed loop block diagrams of a dynamic system where the structural part serves as the main plant of the system and the aerodynamic transfer function as a feedback loop calculated based on the dynamic structural response, it is suitable to carry out the simulation on the platform of Simulink-Matlab. For this purpose the aerodynamic forces have to be conversed in Laplace domain.

The work shows the effectiveness of the flow-structure interactions when the system is considered as a dynamic system where the response can be analysed in time domain and the effects of non-linear factors can be conveniently included simultaneously.

The limit cycle oscillation and stability can be showed numerically by representing the phase portrait of the response. At the speed of airflow below the critical speed of flutter, a constant oscillation may happen due to a free-play nonlinearity. It can be shown that the stability boundary becomes smaller than the critical speed.

Acknowledgements

This research is funded by Ministry of Higher Education Malaysia under Fundamental Research Grant Scheme, no. FRGS/2/2013/TK09/02/1.

Author details

Cosmas Pandit Pagwiwoko[1] and Louis Jezequel[2*]

1 Faculty of Science and Engineering, Department of Mechanical, Materials and Manufacturing Engineering, University of Nottingham Malaysia, Selangor, Malaysia

2 Ecole Centrale de Lyon, Ecully, France

*Address all correspondence to: louis.jezequel@ec-lyon.fr

References

[1] Stanciu V, Stroe G, Andrei IC. Linear models and calculation of aero-elastic flutter. UPB Scientific Bulletin, Series A. 2012;**74**(2)

[2] Trickey ST, Virgin LN, Dowell EH. The stability of limit-cycle oscillations in a nonlinear aeroelastic system. Proceedings of the Royal Society of London. Series A. 2002;**458**:2203-2226

[3] Pereira DA, Vancoscellos RMG, Hajj MR, Marques FD. Effects of combined hardening and free-play nonlinearities on the response of a typical aeroelastic section. Aerospace Science and Technology. 2016;**50**:44-54

[4] http://thehuntformh370.info/sites/default/files/mh370_flaperon_failure_analysis_rev_2.0.pdf

[5] Botez R et al. Aeroservoelastic flutter and frequency response interactions on the CL-704 aircraft. INCAS Bulletin. 2012;(4):41-56

[6] Sinou JJ, Jézéquel L. The influence of damping on the limit cycles for a self-exciting mechanism. Journal of Sound and Vibration. 2007;**304**:875-893

[7] Brunton SL, Rowley CW. Empirical state-space representations for Theodorsen's lift model. Journal of Fluids and Structures. 2012. Submitted for publication

[8] Vio GA, Cooper JE. Limit cycle oscillation prediction for aeroelastic systems with discrete bilinear stiffness. International Journal of Applied Mathematics and Mechanics. 2005;**3**: 100-119